Embryogeny and Phylogeny of the Human Posture 2

Comparative Anatomy and Posture of Animal and Human Set

coordinated by
Djillali Hadjouis

Volume 4

Embryogeny and Phylogeny of the Human Posture 2

A New Glance at the Future of our Species

Anne Dambricourt Malassé

iSTE

WILEY

First published 2021 in Great Britain and the United States by ISTE Ltd and John Wiley & Sons, Inc.

ISTE Ltd
27-37 St George's Road
London SW19 4EU
UK

www.iste.co.uk

John Wiley & Sons, Inc.
111 River Street
Hoboken, NJ 07030
USA

www.wiley.com

Library of Congress Control Number: 2021938413

British Library Cataloguing-in-Publication Data
A CIP record for this book is available from the British Library
ISBN 978-1-78630-729-3

MIX
Paper from responsible sources
FSC® C013604

Contents

1

The 20th Century. A New Science: Human Paleontology

1.1. Introduction

The 19th century saw the recognition of the animal and simian origin of *Homo sapiens*, although not without difficulty, through thought experiments due to the lack of sufficient knowledge in both paleontology and reproductive biology. The long geological periods throughout the history of the Earth and its cooling, as described by Georges Buffon, made it possible to locate fossil and current species according to lineages, or phylums, and then to link these phylums together by looking for "intermediate" fossil species. In this phylogeny of multicellular organisms, which became vertebrates with the horizontal organization of chordates, the primate lineage can be distinguished by the verticalization of the neuraxis starting from the posterior cranial base, the reduction of prognathism and an encephalization that led to *Homo sapiens*, the most verticalized chordate. Verticalization was the guiding thread of the problem, and the bimania (meaning two-hands in Latin) linked to this verticality was then much more significant for anthropologists, along with encephalization, then bipedalism, which was not original, being common in gibbons as well as birds. Whatever one's opinion on the meaning of Man's place in the animal kingdom and the history of the Earth, verticalized anatomy has not left the place assigned to it by the history of animals since Aristotle. It remains the last point on the curve of this complexification/verticalization, which was accompanied by an increase in cognitive capacities.

For prehistorians and anthropologists, *Homo sapiens* is characterized by articulated language and symbolic behaviors manifested by the creation of objects unrelated to primary necessities. These activities concern ornaments, engravings, paintings and sculptures; as innovations that express values, significations, they are bearers of meaning. On the eve of 19th century, the oldest *Homo sapiens* skeletons were still rare, but they were already associated with these productions at prehistoric periods that had nothing in common with the Adamic tradition, still well anchored in the European culture that refused the idea of a simian origin. This filiation was defended at the *Muséum national d'Histoire naturelle* (French National Museum of Natural History, henceforth referred to as the *Muséum*) in 1802 by Jean-Baptiste Lamarck, Etienne Geoffroy Saint-Hilaire and his son Isidore Geoffroy Saint-Hilaire, but 70 years later, it did not find unanimous acceptance with an interesting and strong personality, the geophysicist, physician and anthropologist Armand de Quatrefages.

1.2. Human paleontology, a nascent science

A new prehistoric *Homo sapiens* skeleton was exhumed in 1872 on the lands of the Principality of Monaco, in Cavillon, in one of the Grimaldi caves. Everything pointed to a sepulchral context at levels as old as the Cro-Magnon shelter in Les Eyzies-de-Tayac (France), dating back nearly 30,000 years. The stakes were considerable because the remains carried a mantilla composed of more than 200 shells on its head, covered with red ochre and black hematite. The discovery attested to an already highly ritualized and symbolic relationship with death. The mastery of articulated language left little room for doubt. The excavations, however, did not have any means of guaranteeing its date. Prehistory was not yet a scientific discipline.

Armand de Quatrefages studied the Les Eyzies skeletons with his naturalist assistant, the doctor Ernest Hamy (1842–1902). Together they defined the Cro-Magnon race as "the second fossilized human race", the first being the "Canstadt race" or Neanderthal Man, which was limited to three specimens recognized as: Neanderthal, Gibraltar and La Naulette (see Volume 1). Their study was exocranial, the base of the skull was partially preserved on one of the three adult skulls, but its endocranial surface was not described because of its inaccessibility and there was no question of damaging the unique piece.

New fossils of European Neanderthals enriched the collections in 1886, with three individuals discovered in Belgium in the Spy cave (currently dated at 38,500 years) and especially in Croatia in 1899 in the Krapina cave, with nearly 900 fragmented human remains, nearly 80 individuals dated at 125,000 years old. Traces of anthropophagy would be confirmed.

The great novelty came from Asia with the discovery of fossils of monkeys and great apes, and above all that of a very ancient anthropomorph in Java which generated a great stir among anthropologists. The monkey fossils were collected in the powerful sedimentary deposits of the upper Indus basin called Siwaliks, composed of molasse deposited in the plain by the erosion of the Himalayan range during formation. In the northwestern Indian subcontinent, the deposits were slightly folded and formed the Potwar Plateau. The erosion and scarcity of vegetation gave access to numerous fossiliferous outcrops. Further east, these deposits formed the foothills of the Himalayas, strongly compressed by plate tectonics between India and Asia, sometimes to verticality. The outcrops became long narrow bands covered by tropical forest at low altitude and richer in thorny trees at high altitude. Prospecting conditions were thus limited by vegetation cover and by the small surface area of the fossiliferous layers that inclined rapidly in the depths. The Geological Survey of India (the geological service of the British Empire in India) organized the geological map of the subcontinent. In 1874, Richard Lydekker (1849–1915), a geologist trained at Trinity College, Cambridge (University of Cambridge) was commissioned to map Kashmir, north of the Potwar.

Lyddeker was prospecting the Siwaliks and collecting numerous fossils of vertebrates preserved in the Museum of Calcutta. Among these fossils was a fragment of a monkey's jawbone collected in 1879. Close to the gibbon, the geologist named it *Paleopithecus sivalensis*, then in 1886, he described a canine tooth which resembled that of an orangutan. It was the second fossil of a large anthropoid after the mandible of *Dryopithecus*. Henceforth, the Siwaliks became the cradle of the origin of Man. This perspective was in line with the deductions of Georges Cuvier who thought it was located somewhere in the Himalayan heights, based on paleoclimatic arguments linked to a rise in sea water on a planetary scale. For the high Himalayan peaks, global warming was to result in the melting of the glaciers and the consequent flooding of the plains by the powerful Himalayan rivers.

1.2.1. *The Java erect ape-man, or the missing link,* Pithecanthropus

Haeckel's anthropogeny placed the ancestors of the gibbon in the ancestral lineage of Man and predicted a fossil link that he named *Pithecanthropus*. Since gibbons and orangutans still live in Southeast Asia, Haeckel located the cradle of humanity between Sumatra, Java and Borneo. The probability of finding fossils of this "missing link" in Asia was therefore supported by this maxilla and canine discovered in sub-Himalayan India. Five years later, in 1891, paleoanthropology made a prodigious leap forward in time thanks to a Dutch doctor. Eugène Dubois (1858–1940) had recently defended his thesis in medicine, in 1884, and was soon appointed lecturer with the chair of Anatomy at the University of Amsterdam in 1886. He was neither an anthropologist nor a paleontologist, but he did intend to devote his free time to researching this *Pithecanthropus*. In 1887, he decided to leave Amsterdam to settle in Sumatra with his family, and joined the Royal Dutch Indian Army as a doctor.

Human paleontology was not yet a discipline; the Neanderthal skeleton and the Gibraltar skull had still not been described. Dubois went to the island of Java and prospected the ancient formations of a river, the Solo, on a paleontological site called Trinil. Between 1891 and 1892 many fossils were collected with the help of two engineers and some convicts. These fossils included two human molars and a human-like skull cap with a low vault and a strong bulge above the orbits. Subsequently, a perfectly preserved femur of modern appearance was collected about 15 m from the area. The skull base was missing and only the proximity justified the association of the femur and the skull cap. Dating went back to the Middle Pleistocene (500,000 years in 2020). Dubois published a description of the fossils in Java in 1894 (Dubois 1894), comparing the skull cap with a chimpanzee and decided to attribute the fossil to an extinct genus of Hominids, the *Pithecanthropus*, and to add the femur to define the species as *erectus*, in other words, the erect ape-man. Cranial capacity was estimated at 900 ml, while that of *Homo sapiens* is 1,350 ml on average, that of the orangutan is 320–340 ml and that of the gibbon is 82 ml (White 2007). On the strength of his discovery, Dubois traveled around Europe in 1895 to obtain the approval of anatomists. The monograph was read carefully by the most informed authorities, some 20 experts, and the conclusions were more in the realm of opinions, ranging from a giant gibbon to an extinct genus of intermediate hominin between Ape and Man. The *Société d'Anthropologie de Paris* appreciated the discovery and its implications. In January 1895, Léonce Manouvrier (1850–1927), a specialist in the physiology and anatomy of the human brain, presented his analysis and conclusions (Manouvrier 1895). After seeing the originals, he confirmed his position in 1896 and invited Eugène Dubois to come and present his arguments (Manouvrier 1896). The numerous fossils of vertebrates resembled the most recent species of Siwaliks (2 million to more than 700,000 years old).

If the skull cap was that of a hominin, then the date when Man's ancestors must have appeared would defy theologians. However, the association of the "modern" femur with a skull and a brain both different from that of a *Homo sapiens* posed a problem. The discussion revolved around bipedalism, within the framework of the doctrine of the locomotor origins of human anatomy. Indeed, for Dubois, who apparently did not compare the femur of a man with that of a gibbon, the latter was also bipedal:

> Certainly it was not to be assumed that the femur of an anthropoid becoming bipedal could be very different from a human femur, the function being the same and the form already approximated by the

> origin. (Dubois, *Le Pithecanthropus erectus et l'origine de l'homme*, 1896, author's translation)

Dubois was of course interested in the brain. In 1897, he published a synthesis on the evolution of the central nervous system, which was taken up and translated by Léonce Manouvrier for the *Société d'Anthropologie de Paris*:

> The central nervous system of animals has increased in complexity and volume as their relationships with the external world have multiplied [...]. Mammals have much larger and more sophisticated brains than all other animals and show to some extent, according to the degree of the order to which they belong, an increasing development. They have acquired this superior organization gradually. Marsh[1] and others have shown that the brain dimensions in mammals of the Eocene period are generally smaller than in related forms of the Neo-Tertiary or present day. Placental mammals of all orders in the earliest Eocene terrain have extremely small cranial cavities. In animals belonging to one of these orders, the brain was thinner than the spinal cord, so much so that it could have easily been extracted from the skull through the foramen magnum and the vertebral canal. At the same time, Marsh demonstrates that it was mainly the upper organized parts of the brain, the cerebral hemispheres, which were very poorly developed in these oldest placental mammals, with the result that these forms are even less distant, with regards to this point, from reptiles than recent placental mammals. This progressive development of the brain, in organization and volume, in vertebrates and mainly in mammals, as well as the remarkable volume and complexity in structure that it finally reaches in man, demonstrate that in this organ, the development of function determines the complication of structure and quantity as well. (Dubois, *Sur le rapport du poids de l'encéphale avec la grandeur du corps chez les mammifères*, 1897, author's translation)

The complexification of the mammalian encephalon was maintained during the Tertiary era and Man was therefore the last degree of complexity:

> No animal of the same size has a brain size comparable to that of man. Anthropoid great apes, whose body weight is equal to that of man, do not even reach 1/3, and dogs with the same weight 1/10, with regard

1 Othniel Charles Marsh (1831–1899), nephew of the philanthropist George Peabody (1795–1869) was one of the first American paleontologists. He was the founder of the Yale Peabody Museum of Natural History (Yale University, New Haven, Connecticut).

> to the brain and the weight of a man. Only the elephant, the largest whales and the Rhytine (*Rhytina stelleri*)[2], an animal that disappeared in the last century, outweigh man in terms of the absolute weight of the brain […]. For animals whose body weight is minimal: the relative weight of the shrew's brain is 1/23, the monkey lion 1/26, the Javanese Tupaja 1/41, the mustached bat (*Vesperttlio mystacinus)* 1/42, while the relative weight of the human brain is 1/46 to 1/45: however, among these, in the shrew and the bat, the brain has a much lower organization. (ibid., author's translation)

Encephalization was therefore not a question of absolute mass, volume, or weight relative to the body, it was a question of organization and the duration of growth:

> Not enough attention is paid to the fact that the brain increases less in weight with age than the rest of the body. In humans it has already reached about the same weight as in adults by the ninth year of life, while the body weight increases more than twice as much. The same can be seen, as Dr. Weber demonstrated, in all mammals; the brain has finished growing much earlier than other parts of the body. (ibid., author's translation)

Eugène Dubois did not accept criticism and did not allow access to the fossils before 1923. He was nevertheless supported by Léonce Manouvrier, not on an anatomical basis as there was a lack of comparative material, but on the significant mineralization and the proximity of other bones. Following that, the Java collections were enriched and supported the case for attributing the skull cap, and molars, to at least one extinct genus of hominin. The femur fragments collected confirmed a different anatomy from *Homo sapiens*. The recognition of *Pithecanthropus* by the *Société d'Anthropologie de Paris* and the discovery of a new 30,000-year-old *Homo sapiens* skeleton, buried with evidence of artistic activity dedicated to the dead, were beginning to arouse curiosity. Man's technical intelligence was fascinating, as demonstrated at the 1889 Universal Exhibition with Gustave Eiffel's tower, and the imagination fed by his progress during the Industrial Revolution was not to be outdone by Jules Vernes' imagined voyages to the moon and the deep sea.

1.2.2. *Human paleontology and secularism*

Then, a grave discovered in the depths of a cave revealed the mastery of symbolic language to express immaterial concerns. Spirituality emerged from a process of evolution that singularized the human phylum. The most visionary, and

2 Large marine mammal, belonging to the *dugongidae* family.

therefore the most realistic, understood the issues facing researchers engaged in the study of this process, the emergence of spirituality and the sense of ethics. Consistency is paramount; the brain cell structure is not in itself aware of being a structure with molecules, ions. Natural cell death is one thing, but Man had been able to distinguish between death inflicted on a hunted animal and natural death, developing a sense of the sacred and sacrilege. Ernest Haeckel's anthropogeny was based on a racial scale and served to elaborate projects of societies structured on the principle of reproduction between those better adapted, or more efficient. Albert Gaudry had foreseen the threat of Haeckelian Darwinian monogenism. His anthropogeny prepared the breeding ground for a 20th century that would put them into practice by going as far as the programed elimination of the "maladjusted".

The 20th century saw inherited scientism and social Darwinism, and men drew up historical trials, engraving the recognition of the meaning of the words "crime against humanity" in law and collective memory (London Agreement of August 8, 1945, Statute of the International Military Tribunal).

When ontology disappeared from thought in favor of a scientist and materialist ideology, in the sense of the reduction of all that is thought, to blind physico-chemical determinisms, the psyche no longer had a basis to justify the difference between a being defended by rights, on the one hand, and biology, the object of experimentation in the laboratories programing the killing of the individual, on the other hand.

Death is as sacred as life, the human reflexive consciousness is not insensitive to it, and paleoanthropologists now had to seize this sensitivity. Prehistorians understood this. The human singularity manifests itself through the creation of meaningful expressions, and thus through symbols, a kind of transcendence without which forms and movements are neither signified nor significant, and this transcendence also applies to Man, whose reflexive consciousness calls itself "human" and covers its body with symbolic adornments.

Prince Albert 1st of Monaco (1848–1922) was aware of the importance of the heritage of the Grimaldi Caves, which were enriched with new Cro-Magnon burials. The prince followed the courses by the chairs of the *Muséum*, in particular in paleontology with Albert Gaudry. In 1895, he took the decision to personally supervise the excavations in the sectors of the caves still intact according to his own method. This method consisted of stratigraphic surveys of the archeological and faunal furnishings with their spatial coordinates. In 1897, advised by Armand de Quatrefages, he entrusted the scientific supervision to a French team composed of a young geologist attached to the paleontology chair of the *Muséum*, Marcellin Boule, the Toulouse prehistorian Emile Cartailhac (1845–1921) and the anthropologist René Verneau (1852–1938). The latter had been asked to bring back skeletons from

Guanches (Canary Islands), similar to the Cro-Magnon race, for the *Muséum*'s collections. The Prince's excavations were a success. In 1901, two new skeletons, one adult and one child, were discovered in a stratigraphic context, confirming the Upper Paleolithic age of the tomb of Cavillon and going back to the Gravettian culture (28,000 years). On the strength of this immense success, the Prince created the *Musée d'Anthropologie préhistorique de Monaco*.

René Verneau convinced him to organize the 13th International Congress of Prehistoric Anthropology and Archaeology in Monaco. The *Muséum national d'Histoire naturelle* was represented by Marcellin Boule. The Prince accepted and the congress took place at the Palais de la Mer in April 1906:

> Today the *Palais de la Mer* welcomes Anthropology [...]. The progress of Maritime Biology and Zoology allows us to use the revelations of Paleontology to constitute the scale of the infinitely numerous transformations through which a force, which we call life, has caused organic matter to pass. And Meteorology, so intimately linked with Oceanography by ceaseless relationships, helps us understand the fluctuations, migrations and geographical distribution of beings, including those of Man. (Albert Grimaldi, Prince of Monaco, *Discours d'ouverture*, 1906, author's translation)

Marcellin Boule (1861–1942) was born in Auvergne, in the volcanic *Massif Central* of France. He was fascinated by geology from his adolescence. He studied medicine at the University of Toulouse, and that is where he met Emile Carthailhac, who introduced him to paleontology and prehistoric excavations. A scholarship enabled him to travel to Paris to attend classes at the *Collège de France* given by the *Muséum*'s professors, and that is how he met Albert Gaudry in 1886. Gaudry advised him to study geology applied to paleontology and entrusted him with a thesis entitled "*Essai de Paléontologie stratigraphique de l'Homme*" (Essay on the stratigraphic paleontology of man). It was a first that would make history and serve as a methodological reference. After this training in Paris, he returned to his native regions to pursue a thesis on the volcanoes of Auvergne at the University of Clermont-Ferrand. The University of Montpellier offered him the chair of Paleontology, but he chose to join Albert Gaudry at the *Muséum*, who immediately entrusted him with the organization of the new gallery of Paleontology in collaboration with Ernest Hamy (Armand de Quatrefages chair), while the gallery of Comparative Anatomy was managed by Paul Gervais' successor, Georges Pouchet (1833–1894) (see Volume 1). The gallery was inaugurated with great pomp and circumstance in 1898 by the President of the Republic. Albert Gaudry left his chair in 1902, Marcellin Boule was elected professor of the *Muséum* and holder of the chair of Paleontology in 1903. The geologist had been able to judge the state of fossilization of the bones collected in Java by Eugène Dubois when he came to

Paris. He did not adhere to the Pithecanthropic thesis; for him, the skull cap was that of a very large gibbon. A secular republican, Boule was committed to Lamarck's Transformism and conscious of the reductionism of the Darwinist doctrine, as defined by Haeckel.

From July 1, 1901, a French law prohibited associations from providing education without the agreement of the State, and religious congregations were forced to flee abroad. This was particularly the case for the Jesuits, with their apostolic schools, novitiates and scholasticates, who found refuge in Hastings (Great Britain), Jersey, Monaco, Belgium, Spain, Italy and Holland.

Abbot Henri Breuil entered the *Muséum*'s sphere of influence during his studies at the *Grand Séminaire* of Issy-les-Moulineaux in 1895, during which he discovered the theory of evolution thanks to an abbot, Jean Guibert, professor of natural sciences and friend of Albert Gaudry. Breuil discovered a passion for prehistoric art. In 1902, Émile Cartailhac invited him to study the cave paintings in the caves of Marsoulas (Pyrenees) and Altamira (Spain). Henri Breuil understood how the animal *Homo sapiens* was a living being different from other animals, at a time when France was going through a violent anticlerical crisis that sought to control the national education system, and thus the colleges of religious congregations, as well as the major seminaries (Hurel 2011). On December 9, 1905, the law on the separation of Church and State was passed. It proclaimed freedom of conscience: the State was neutral, religious doctrines were taught in the free schools of the congregations, they were no longer to intervene in the field of scientific research, hard sciences and humanities financed by the State.

Five years after the election of Marcellin Boule, a national event, but one of international scope, marked human paleontology with the discovery of the oldest burial site. It was 60,000 years old and it was not that of a Cro-Magnon Man, but that of a Neanderthal.

1.2.3. *The first Neanderthal Man in French territory and his ancestor in Germanic lands*

In August 1908, France discovered its first skeleton of Neanderthal Man in southwest Corrèze, in a small cave in the commune of La Chapelle-aux-Saints. The skeleton was of an adult, it was articulated, and therefore had not moved since its burial. The discoverers were the three Bouyssonie brothers, Amédée (1867–1958), Jean (1877–1965) and Paul (1887–1972). It was almost complete and analyses of the burial conditions confirmed its intentional origin. The *Muséum de Paris* purchased the fossil and added it to the paleontology collections, rather than the anthropology collections. The skeleton was therefore under the responsibility of Marcellin Boule. A Neanderthal could finally be compared to the Cro-Magnon Men.

At the beginning of the 20th century, the *Muséum* became the leading research establishment in Europe because of the importance of its collections and the organization of its teaching and development with 16 chairs. The research of the extinct anatomical stages between fossilized monkeys and *Homo sapiens* had finally found its institutional basis. Study of the Neanderthal skeleton was entrusted to the chair of paleontology, at the head of the largest fossil collection in the world. This choice formalized the birth of evolutionary human paleontology as an academic discipline. It was a consecration to the memory of Jean-Baptiste Lamarck who had dared to imagine a century earlier, in 1802, a mere 100 m away, the inconceivable filiation of present-day apes and Man from a line of common ancestors. It was recognition for the posterity of Etienne and Isidore Geoffroy Saint-Hilaire, who for years fought against the dogmatism of Georges Cuvier and his many disciples. It was an immense gratification for Albert Gaudry, who carried the gallery of Paleontology and Comparative Anatomy at arm's length, both seen as second-order disciplines, classifiers, with no interest in the real concerns of the human condition.

Gaudry died on November 27, 1908, a few days before the presentation of the skeleton from La Chapelle-aux-Saints to the *Académie des sciences* by the director of the *Muséum* on December 14, 1908. From now on, *Homo sapiens* could set off in search of its anatomical, cultural and spiritual origins, in the long history of the Earth, discovering very old rituals celebrating the departure of a deceased person, whether Sapiens or Neanderthal. The Biblical Genesis was nothing more than a collection of texts written by humans who disappeared more than 2,000 years ago, one moment among many, in the continuity of distant human lineages where the questioning of the meaning of gestures was already consciously manifested, up to the killing of the hunted animal and the human, sometimes consumed. Prehistory was the true book of memory that kept the traces of the emergence of consciousness and its search for meaning. Its archives confirmed the singularity of humanization with the need to signify and to signify oneself, from conception to death, the need to name in order to make the collective understanding of a common reality in permanent interaction with the world intelligible.

While France discovered its Neanderthal, a German industrialist with a passion for the origins of Man, Otto Schoetensack (1850–1912), obtained a Hominid mandible in a sandpit that he had carefully monitored near the village of Mauer in the Heidelberg region. The jaw was short and broad, massive, with a low mandibular ramus that was very extensive on the surface, corresponding to a powerful masseter (masticator) muscle and a pronounced prognathic symphysis that excluded it from the *Homo sapiens* morphology (Schoetensack 1908). Its stratigraphic position was well identified and gave it a much older age than the Neanderthal of

La Chapelle-aux-Saints. Schoetensack named the fossil *Homo heidelbergensis*. Its recent dating is 600,000 years old, that of La Chapelle-aux-Saints is 60,000 years old. Prince Albert 1st of Monaco visited Marcellin Boule in his paleontology laboratory in Paris. Prehistory and human paleontology did not have a research center. The materials from the excavations, both archeological (tools carved in stone, bone and ornaments) and fauna, were stored in the museums and there was no specialized study laboratory, and consequently, no teaching chair. It was then that a large-scale project took shape with the Prince and Henri Breuil, supported by Marcellin Boule. Animal fossils had their cathedral in the *Jardin des Plantes*, human fossils would have their palace outside the walls of the Garden, thanks to a foundation offered by the Prince under French law and recognized as being of public utility.

In France, a foundation is "a patronage organization that makes a patrimony available to serve a cause of general interest. It is created by decree in the Council of State, after the request has been examined by the services of the Ministry of the Interior".

1.2.4. *The* Institut de Paléontologie Humaine, *a foundation of Prince Albert 1st of Monaco*

The *Institut de Paléontologie Humaine* (Institute of Human Paleontology) was created on July 23, 1910 with the personal funds of Prince Albert 1st of Monaco. The foundation owed much to exchanges with Henri Breuil. Completed in 1914, the building is a veritable palace built over five levels with a floor area of 1,200 m^2. The three facades are decorated with friezes that catch the eye of the passer-by, illustrating different cultures, from different latitudes, Inuit, native Americans, central Africans, converging toward the monumental entrance door, which takes up the style of glass and wrought iron of the great gallery of paleontology.

The skull of the La Chapelle-aux-Saints Man is enthroned above the porch, while on either side, an artist has carved a profile of a buffalo or sculpted the body of a naked woman in the rock (the Venus of Laussel, in the Dordogne, dating to the Gravettian period). The skull of the Cro-Magnon Man is on the left, that of La Chapelle-aux-Saints on the right, associated with prehistoric ornaments. Part of the coat of arms of the Principality of Monaco overlooks the monumental entrance, but the two monks and the motto "Deo Juvante" ("with God's help") have been removed to mark the secular spirit of the place. By climbing the steps to the porch, the visitor understands that here, Man is studied not only in his animality, but also with what makes him a singular being in the evolution of the animal kingdom, through the expression of a symbolic thought that is revealed through the creative act of the artistic gesture.

The architecture is organized around a conference and exhibition room under a huge glass dome, whose height occupies the first three levels, a large library, three rooms for the collections corresponding to the three teaching chairs: human paleontology, fauna and prehistoric ethnology, plus photography and chemistry laboratories, as well as a drawing room and work offices. Marcellin Boule was the director-administrator, Henri Breuil held the chair of prehistoric ethnology, René Verneau held the chair of prehistoric anthropology and Hugo Obermaier (1877–1946) held the chair of geology applied to prehistory. The First World War postponed the inauguration of the foundation by the President of the Republic until December 23, 1920, in the presence of Prince Albert 1st:

> It is to help anthropology overcome the barriers that separate it from the complete truth that I founded the Institute of Human Paleontology, giving it all the necessary independence to lead our spirit towards the light. And I entrust its interests to men who serve Science with a sincerity capable of developing its strength and protecting its march against the influence of passionate interventions. (ibid., author's translation)

Marcellin Boule was now the director of the Paleontology Laboratory of the *Muséum* and the *Institut de Paléontologie Humaine* (IPH). He was at the head of the largest scientific organization covering the study of the evolution of fossil species, with collections, laboratories and teaching chairs. The paleoanthropological collections were rapidly enriched from 1909 to 1921, following the exceptional discovery of Neanderthal burials in the Ferrassie Cave in the Dordogne (southwestern France), containing two adults, three children, a newborn and two fetuses. The IPH thus became the first research center in the world entirely dedicated to the study and understanding of the origins of the human being.

Boule published his imposing monograph on the skeleton of La Chapelle-aux-Saints between 1911 and 1913 (Boule 1911–1913). At the same time, Henri Breuil transmitted the evolutionary doctrine of the chair of paleontology in his courses at the Institute:

> In order to see all these forms come out of a single origin, we have to suppose three things: humanity began in a single region, probably towards the East, the human body was able to transform, the first men dispersed, living in physical and moral conditions that profoundly modified their organisms. (Breuil, in Hurel (2011), author's translation)

1.3. Presuppositions

1.3.1. *The Neanderthal Man, a stooped posture?*

The degree of axial straightening of the Neanderthal can finally be compared to *Homo sapiens*. The base of the skull was not complete, but it was sufficient to compare the angles of verticalization, with the vertebrae and sacrum. The frontal curvature seemed to be lower than in *Homo sapiens* because of the particularly pneumatized ("inflated") supraorbital torus above the orbits, the adult facial block was in front of the frontal lobes and not below them as in the Cro-Magnon Man. The brain was more voluminous, but the impressions of the lobes were different and the prefrontal cortex was especially less developed. The cerebellar lobe was clearly higher than that of the Cro-Magnon Man, it was also smaller in absolute value and in relative value to the brain, which corresponded to a less verticalized clivus and a more open sphenoidal angle. The face was prognathic, the mandible had no chin. The geometric protocols in the median sagittal plane confirmed a general straightening movement, but it was not as pronounced as in *Homo sapiens* (Figures 13 and 14 in Boule (1911–1913)). The reconstitution of the profile of the spine with isolated vertebrae did not show cervical and lumbar lordosis (forward curvature), so it was not sinusoidal like *Homo sapiens*. The femurs were curved, more stocky.

The curvature of the femurs gave the impression of a body in a forward bent position, which was accentuated on the La Chapelle-aux-Saints Man due to a congenital dislocation of the left hip, but which really did exist and was systematically found in Neanderthal patients. The general conformation of the skull with high neck and facial prognathism, as well as the slightly flexed profile of the thighs, gave a general cranio-spinal appearance intermediate between that of the great ape and *Homo sapiens*. Neanderthal Man thus appeared to be less vertical and less cerebralized than the Cro-Magnon Man. The appearance reflected an anatomy and cerebral development of an evolutionary stage preceding that of *Homo sapiens*. However, Boule did not venture to interpret the brain's imprints in terms of cognitive capacities beyond what would be possible with analogies with pathological conditions of current *Homo sapiens*.

1.3.2. *Embryologists, geneticists and paleontology at the beginning of the 20th century*

Molecular genetics would become a game-changer, but it was still barely studied. Its development would displace comparative embryology from paleontology, and would allow scholars to reason on the systemic and dynamic scale of the organism, and thus reflect on evolutionary processes, with an opening to complex systems science from gametogenesis. Disciplines would not lose sight of morphogenesis with biomathematics and thermodynamics, they would make it possible to preserve the scientific method of the study of the processes of the evolution of forms, the theoretical and thus mathematical formalization of the changes in dynamic trajectories that set up the structures and organs.

Wilhelm Johannsen (1857–1927) was a Danish botanist. In 1909, he created the word *gene* as opposed to pangene or "gemmules" from Charles Darwin (see Volume 1). The gene is a discrete unit. In 1911, he proposed the terms genotype for germ cell lines, and phenotype for soma cell lines. Mendel's laws, which had gone unnoticed since 1866, were rediscovered in botany and independently by the Dutchman Hugo de Vries (1848–1935), who created the word mutatio, the German Carl Correns (1864–1933) and the Austrian Erich von Tschermak (1871–1962). In France, genetics was very slow to develop. Lucien Cuénot (1866–1951) was one of the very few geneticists at the French University. Professor of Zoology since 1898 at the Faculty of Sciences of Nancy, his research took up Mendel's laws and studied them in mammals for the first time, with crosses of gray and albino mice. In 1902, he confirmed the law of transmission for the variability of a characteristic, which was common to both plants and animals. Cuénot then became one of the main actors in the development of genetics.

The nucleus of cells contains macromolecules called chromosomes, themselves composed of these discrete entities called genes, but their behavior during reproduction remained a black box. The first experimental studies aimed to produce mutations, and were carried out in 1908 by the American geneticist Herman Müller (1890–1967). He had the idea of exposing drosophila to X-rays and making them reproduce with unexposed drosophila. This invisible light was discovered at the University of Würzburg in November 1895 by the physicist Wilhelm Röntgen (1845–1923). The results were convincing since some drosophilas were born with malformed wings. X-rays made it possible to infer not only a role of sex cells in heredity, but also to understand that the memory of heredity was on the scale of X-rays, in other words, on the scale of the bonds between molecules, the scale of the forces of attraction that make up these radiations. Physicists did not yet know the structure of chromosomes, nor the nature of X-rays. These were photons, electromagnetic waves. In the future it would be confirmed that it is the segments of chromosomal macromolecules that are altered.

The discovery of hereditary memory on the scale of electronic links brought back the great questions about generation that René Descartes himself pondered, incredulous about the infinite dimension beyond the smallest atoms, as well as Isaac Newton, for whom the beginning of any organization, whatever the scale of space-time, was a question of forces of attraction.

The new doctrine of generation was Hugo de Vries' "mutationism", which no longer concerned the organism, but the gene. Genetics gained momentum from the 1920s onward with Ronald Fisher (1890–1962), John Haldane (1892–1964) and Sewall Wright (1889–1988). If mutations in these discrete entities, or genes, had no causal link with an adaptation constrained by the environment, then the Lamarcko-Darwinian doctrine would collapse. If X-ray experiments were to be considered as the model of evolution, the primary cause would be a simple random perturbation of the duplication of electronic bonds between molecules forming chromosomes. The evolution of embryogenesis would be reduced to a sequence of genetic copying errors, which was not consistent with observations in paleontology and comparative anatomy. Flies without wings or with malformed wings are unfit to fly, in nature they starve and do not transmit genetic copy errors, they are abortive.

The mutationist model of the drosophila did not correspond to the level of embryonic organization of the fly, since the fly is not transformed. These mutations did not correspond to the problem of the neuroaxis straightening, because simian embryogenesis was not a variety of the embryogenesis of a prosimian, it was its reorganization with a complexification of the central nervous system.

The German embryologist Hans Spemann (1869–1941) thus resumed van Baer's research on vertebrate embryogenesis since gastrulation. A Doctor of Medicine at the University of Heidelberg, he continued his research on the embryogenesis of batrachians, in order to isolate and better identify the mechanisms that preside over the differentiation of the three embryonic layers: the ectoderm (future nervous system and part of the dental germs), the mesoderm (from which the dorsal cord is formed) and the endoderm. In 1924, thanks to cell transplants, he revealed a very early territory in the cell cluster derived from the blastula, which he named the "organizing center" with mechanisms qualified as "induction". The Nobel Prize in Physiology or Medicine was awarded to him in 1935 "for his discovery of the organizing effect in the development of embryos". Such was the state of the question on the generation of the organizational plan of a species, when human paleontology and prehistory took off. That said, Neanderthal embryogenesis was not a subject for reflection and the orientation of its dorsal chord was not even conceptualized.

2

Asia, The Cradle of Humanity

2.1. Teilhard de Chardin, a destiny from Piltdown to the *Muséum*

Marcellin Boule was in contact with French naturalists abroad. France had enjoyed concessions in China since 1860, ceded by the imperial power of the Qing Dynasty, notably in Tien-Tsin (Tianjin) in northeast China. It was a port city on the Yellow Sea, 100 km south of Beijing. Father Emile Licent (1876–1952), a naturalist from the French Jesuit mission, had taught there since 1914. He regularly went prospecting in northern China: the Tcheu-Ly plain, the Mongolian plains, the Yellow River basin or Hoang Ho, the province of Shaanxi and the Tibetan steps in Gansu. The year 1920 was promising. Licent came back from Gansu with 84 camels loaded with fossils collected in loess, as well as quartz stones that appeared to be cut. They were collected at the base of these same loess that marked the end of the Middle Pleistocene (120,000 years). Fossils were piling up at the Mission. Licent needed a museum, a laboratory and a collaborator to sort, identify and store them. He therefore contacted Marcellin Boule and sent him specimens, who entrusted them to a paleontologist who had just defended a particularly brilliant thesis, Pierre Teilhard de Chardin (1881–1955), also a Jesuit, like his colleague from Tien-Tsin.

Pierre Teilhard de Chardin was a native of the Auvergne, and the history of the Earth had fascinated him since childhood, especially geophysics and tectonics, which he studied on the island of Jersey. He was also passionate about paleontology, which he practiced by collecting fossils, particularly in Fayoum, Egypt, when he taught at the School of the Holy Family in Cairo. He passed his theologate in 1911, in exile at Ore Place, Hastings (UK). When his studies were complete, he returned to France at the age of 31 years. On July 19, 1912, he went to the paleontology laboratory of the *Muséum*, asking to meet Marcellin Boule who was about to leave for his summer vacation. The latter received him straight away:

> Do you remember our first interview, around mid-July 1912? That day, I came timidly, around 2 o'clock, to ring the doorbell. It was the (sacred!) eve of your departure for your vacation and very busy. However Thévenin relented. You received me anyway. And you suggested to me to come and work at your home, at Gaudry's school, at your school. (Teilhard de Chardin, *Jubilé de Marcellin Boule*, 1937, author's translation)

Since the discovery of L'Homme de la Chapelle-aux-Saints, Marcellin Boule had become the French authority on human paleontology. The reason for this warm reception could only be a strong argument. It was a still unheard of discovery in the south of England, in the village of Piltdown, 100 km from Hastings. Teilhard witnessed it on June 2, 1912, invited to observe by the paleontologist Sir Arthur Smith Woodward (1864–1944). This discovery firmly placed the cradle of *Homo sapiens* in England and left the Neanderthals to old Europe.

Teilhard had read *On the Origin of Species* by Darwin but he did not know Lamarck, he had no training in anthropology and prehistory, he did not know what an excavation was and had no interest in the origins of Man. Marcellin Boule took the news seriously and directed Teilhard de Chardin to Henri Breuil in order to learn how to excavate and identify stone tools. He directed him to the work of the Paleontology Laboratory as early as 1912. Boule began the study of the Neanderthals of La Ferrassie; nobody understood the co-evolution of the neurocranium and the face better than him.

The discovery of Piltdown was announced by the British press in December of the same year and an illustrated description was published in 1913 (Dawson and Smith Woodward 1913). Boule immediately concluded that it was an ape's jaw – which can be seen at first glance – and fragments of a human skull, with the "modern" cerebellar fossa. It was a typical model of the *Pithecanthropus*, but inverted: a modern neurocranium with a simian face. In other words, this showed the degradation of the fundamentals laid down by Georges Cuvier, with the principle of correlations. Boule never included these bones in his encyclopedia on human fossils and Teilhard would hardly ever refer to them, relying on his master. The "fossil" was a fake, composed of a neurocranium of present-day man and an orangutan mandible; the deception would be recognized definitively in 1959.

The Piltdown Man was the most incredible fraud in paleoanthropology. Teilhard was 29 years old and finishing his theology studies. He would never get over the idea that Sir Arthur Smith Woodward had deceived him, because his presence brought no credit, he represented no institution and he wasn't looking for any kind of future in prehistory. The author of the scoop, Charles Dawson, had specified in his article, co-authored with Smith Woordward, that Teilhard was not involved in the discovery, the glory belonged to Charles Dawson alone, a subject of His Majesty King George V. In his book on the Piltdown affair, Herbert Thomas, a paleontologist affiliated to the chair of Prehistory and Paleoanthropology at the *Collège de France*, did not hesitate to qualify the accusations of fraud brought against Teilhard by the British paleontologist Louis Leakey (1903–1972) and the American paleontologist Stephen Jay Gould (1941–2002) as being in bad faith (Gould 1980). The investigation was prefaced by Yves Coppens, honorary member of the Teilhard de Chardin Foundation, located in the *Jardin des Plantes* of the *Muséum national d'Histoire naturelle* (Thomas 2002).

Welcomed by Marcellin Boule, Pierre Teilhard enrolled at *La Sorbonne* in 1912 to study for a bachelor's degree in geology and paleontology. In 1914, Boule entrusted him with the study of an important collection of carnivore fossils from the region of Reims, bequeathed to him by Victor Lemoine, Professor of natural history at the medical school of said city (and future director of the *Muséum*). These fossils were among the oldest specimens of mammals, between 59 and 49 Ma, from the beginning of the Eocene (the second stage of the Tertiary era). They were collected in the northern hemisphere, in an age when tropical forests extended beyond 50° latitude, from Canada to Mongolia. This was the warmest period since the end of the Secondary era (the extinction of the dinosaurs).

French research had lagged behind in the comparative study of the fauna of this period.

On August 3, 1914, Germany declared war on France, and Teilhard was mobilized as a stretcher-bearer and a nurse, sent to the front and moved to the trenches of the various large battlefields. He came back deeply affected by the experience. When the First World War ended, he obtained his bachelor's degree in 1919 and enrolled for a thesis in 1920. Teilhard was confronted with scientific knowledge that had discredited the supernatural character of the Bible and had put the character of Yeshua (Jesus) back into its historical context by Ernest Renan (1823–1892), a convinced evolutionist. The Jesuit had to reconsider 30 years of Catholic doctrine, shaken by 4 years of war, searching for the wounded between shell craters and corpses. He followed the courses of geologist Gustave Haugh (1861–1927) in the tradition of Georges Buffon at the Sorbonne, and completed his study of the tectonics of the island of Jersey, learning to reason on a planetary scale:

> It is rational to admit that each continental unit has had its own fauna as long as it has remained isolated, that migrations occur whenever land communications are established with a neighboring unit, and that the fauna best organized in the struggle for existence manages to establish itself and push indigenous fauna to the more remote parts of the continent. If, later, continental unity is fragmented by a partial

> collapse, the remnants of fauna will allow the old connection to be re-established through thought. We can see what valuable services the study of terrestrial fauna can render us in attempts to reconstitute fragmented continents. (Haug, *Les géosynclinaux et les aires continentales*, 1900, author's translation)

2.2. The primate ancestor of the human lineage in Montauban

Marcellin Boule suggested to Teilhard that he complete the Lemoine collection with those of the Museum of Montauban (in southwestern France) where fossils of carnivores from the phosphorites of Quercy dated from 38 to 34 Ma (Terminal Eocene) are preserved. Among the species of these phosphorites are tiny primates barely the size of a mouse, first described by Henri Filhol (1843–1902), the son of the curator of the Museum of Toulouse. They were classified in the group of prosimians of the Tertiary era called *Necrolemur*. Henri Filhol ended his career at the *Muséum national d'Histoire naturelle* as chair holder of Comparative Anatomy from 1894 to 1902. Paleontologists would therefore search among these fossils for those that would allow the identification of a possible ancestral lineage of the simians. In 1920, the name prosimian was still being used; it included several current families known in equatorial and tropical Africa, as well as in Madagascar and Southeast Asia. Lorisidae (from the name Loris described in 1796 by Etienne Geoffroy Saint-Hilaire) lived in Africa and Asia; Galagidae (from the name Galago, also described in 1796 by Geoffroy Saint-Hilaire) lived only in Africa; Lemuridae (lemurs) were confined to the island of Madagascar; and finally Tarsiidae (the small tarsier that fits in the palm of the hand) were limited to Southeast Asia. Paul Gervais had decided in 1854 to separate the lemurids and the tarsiids by bringing the latter closer to the simians.

The search for the ancestors of the simians thus focused on those of the tarsiids and this problem was the second subject of Teilhard's thesis. Fossils of prosimians were accumulating, as well as names and the need to clarify their nomenclature. In North America, there were, for example, *Omomys* (1869), *Notharctus* (1870) and *Anaptomorphus* (1882). Similar species were collected in South America. Others in Europe included Cuvier's famous *Adapis* (1821), *Microchoerus* (1846) and Filhol's *Necrolemur*. These fossils and the present families gave two main phyletic orientations because of the external appearance of the skull of the tarsier. All of the skulls had a long muzzle aligned in the extension of the spine, except the tarsier, which gave the impression of a globular neurocranium, a very short face and a great occipital hole (foramen magnum) tilted forward and downward. The approach to simians was built on these adult exocranial appearances, without considering their embryonic morphogenetic trajectories.

The internal base and the sphenoidal angle were ignored because observation and measurements were not conceivable; they would destroy the integrity of the fossils, which were rare. As for Broca's measuring instruments, they could only be used for very large skulls. Therefore, the sphenoidal angle and the straightening of the base would not be studied by paleontologists and would not enter into phylogenetic criteria. The logic of comparative embryonic chronology, which should have developed after Haeckel, was not to have a future in paleontology. Today, nomenclators (classifiers) no longer speak of prosimians, and they distinguish at least two lineages, one that brings together species with long snouts and small orbits, the other that brings together species with very large orbits and seemingly have short snouts. The first are the Strepsirrhini (e.g. Cuvier's *Adapis* and lemurs), the second are the Haplorrhini with the tarsier and a fossil species like *Omomys* (North America). The nomenclators then explained that the simians were also Haplorrhini. In other words, the classifiers invented a category based only on adult exocranial features, without being interested in the orientation of the clivus, which was not taken into consideration. The origin of its straightening was left aside by paleontologists and therefore ignored, only the orientation of the occipital hole counted, as in Daubenton's study.

However, these analogies with simians were only made due to the disproportionate development of the orbital cavities; the skull disappeared under two enormous eyes, while in the median sagittal section the face was no less prognathic than a *Loris* and the primordial cranium (the base) was flat. It is the development of the embryonic optic capsules that disorganizes the shape of the forehead and thus gave the impression that there is no muzzle (see Chapter 5). These differences have nothing in common with the embryonic mechanisms that straighten the dorsal cord and separate the simians, or Simiiformes, from the prosimians, which in fact include Tarsiiformes and their ancestors.

If Georges Cuvier were still alive, he would have adopted the same posture as that of Etienne Geoffroy Saint-Hilaire who could not distinguish between analogy and homology. Consequently, as we have seen, the latter did not conceive of the need to take into account, in the nomenclature, the primary reality of the organizational plans structured from embryogenesis. Yet, these plans did not fit into each other, they emerged from each other. Georges Buffon would have said that the Haplorrhini category was a view of the mind, that it had no basis in reality. There was no cranio-spinal embryogenesis common to tarsiers and simians.

2.3. Teilhard de Chardin sets out the main phylogenetic principles

As long as the embryonic organization of primates was not established, the place of tarsiers in human phylogenesis would continue to divide nomenclators. Teilhard considered that he had solved this primordial question with his thesis. He cleared a fragment of mandible and skull (maxilla and orbits) from a gangue of sediment preserved in Montauban. These remains resembled the tarsier and no such specimen was more complete: the rare fossils were fragmented. The size was very small in comparison, but the fossil resembled it with the oversized development of the orbits. Teilhard then made the very first comparison between Eocene primates from North America and Europe (see the geological scale, Figure 2.1). He concluded his study by considering this fossil a representative of the tarsier lineage and he put forward a reasoning that would mark the Teilhardian phylogenetic logic.

Fossils are discovered much later than the emergence of the type, because it takes time for generations to reproduce, to expand geographically before reaching a sufficient population density for the probability of finding a fossil to begin to be relevant. In other words, and this is a matter of common sense, the collected fossilized species are already specialized and not generalized and confused with the ancestral morphotype. As a consequence, the probability of finding the earliest example of a stem population is almost zero, so much so that Teilhard called this the "blank of birth" in the paleontological archives of a new lineage.

Teilhard claimed his conclusions without waiting for his thesis. He published them as early as 1921 in the "*Annales de paléontologie*" founded by Marcellin Boule and presented to the *Institut français d'Anthropologie* (French Institute of Anthropology) on March 16, 1921, under the presidency of Boule, who gave him the full weight of his authority. This Institute was a learned society founded in 1910 on the initiative of the ethnologist Paul Rivet:

> The question arises as to whether the Tarsiidae (in the new sense) should not be more completely removed from the group of lemuroid forms, to make a sub-order between Lemuroidea and simians. Their line of evolution now appears so defined, so independent of that of the other Primates, that one is tempted to give them this exceptional place. But then one sees, once again, how much we must project far back in the past, the date of the dispersion of the Primates […]. The small Malaysian Tarsier, which zoologists did not know whether they should designate as a Lemur or a Monkey for a long time, appears decidedly today as an independent and isolated type in nature; it is the only survivor of an important group which covered North America and Europe before the Oligocene. (Teilhard de Chardin, *Sur l'origine tarsienne de l'Homme*, 1921a, author's translation)

This conclusion is today accepted by the international community. The oldest lineages of primates dating back to the Eocene can be divided into two phylums: on the one hand, that of the current families of Lorisidae, Galagidae and Lemuridae, united in Strepsirrhini, and, on the other hand, that of Tarsiidae. Both phylums derive from an even older primate population that dates back at least to the Paleocene (59 million years ago) (see geological time scale, Figure 2.1).

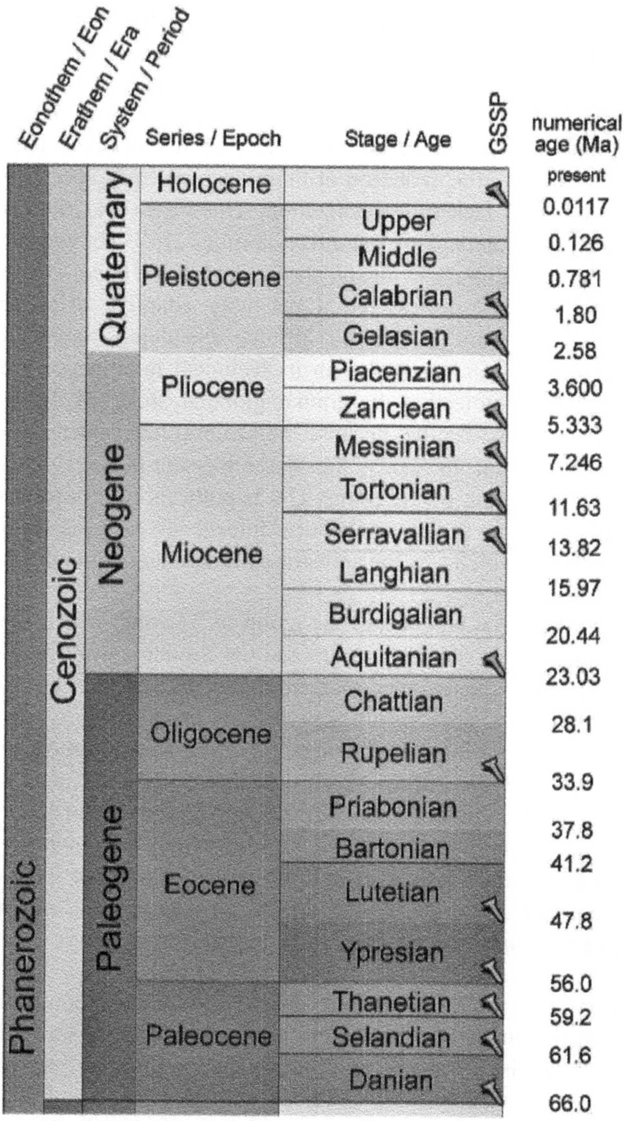

Figure 2.1. *Geological time scale of Tertiary and Quaternary eras, International chronostratigraphic chart 2017 (source: Institut Cartogràfic i Geològic de Catalunya). For a color version of this figure, see www.iste.co.uk/dambricourt/embryogeny.zip*

Teilhard continued his reasoning: first, the fossil families of Strepsirrhini were already all specialized, and, second, the tarsier type had been stabilized since the Eocene (35 million years ago); thus, there was necessarily a third non-specialized lineage able to transform into Simiiformes. The species of this lineage were said to be "generalized". This notion of generalized lineage at the origin of the lineage of simians (Simiiformes) was absolutely crucial, and it would be taken up and developed by Teilhard's successors, such as Jean Piveteau.

Hominization was possible because species did not specialize. The criterion that Teilhard used to identify the evolution of these "generalized" species was the organ that compensated for a lack of specialization. This organ was the central nervous system with the development of cognitive abilities. In other words, the guiding thread was encephalization. Obviously, the reference is *Homo sapiens*, which did not have specialized subspecies, but a plasticity adaptive to all environments, including the most extreme in latitude, altitude and depth. Sapiens did not need wings, fins or gills, it compensated with its technical intelligence. Later, Teilhard would be surprised to note that the cranial globular shape of the tarsier did not correspond to an encephalization, because the cerebral hemispheres were as smooth as those of a Strepsirrhini. In reality, this shape was not due to the brain but to the exuberant lateral extension of the orbits. His hypothesis of a third Eocene phylum prompted him to conceive of a simiiforme lineage as early as the Eocene (thus before the Oligocene, which saw the appearance of simians or monkeys):

> If, even before the Oligocene, the group of Tarsiers, which seems to be interposed between the Lemurs and the Simians, was approaching maturity, it must be assumed that, from that time on, the group of the superior primates[1] was in full growth. This gives depth to our views on the group to which we belong: a Tarsier found, almost entirely formed, in the Phosphorites may mean that, somewhere on Earth, at that same time, Anthropoids were already sketched out [...]. They come, independently of each other, from a still unknown group of very small animals with a large brain which must have lived in the Paleocene, or even an earlier epoch. (Teilhard de Chardin, *Sur l'origine tarsienne de l'Homme,* 1921, author's translation)

This theory of a third phylum of large-brained Simiiformes, already differentiated at the end of the Secondary era, was the reasoning of Teilhard with two evolutionary levels, that of the trunk or the succesion of morphotypes illustrated by generalized (not specialized) species, and that of the branches, or the specialization of species which have diverged from the trunk. The trunk was the dynamic of encephalization, while the divergences were the specializations that

1 *Singes supérieurs* in French.

stabilized the organism and slowed down the complexification dynamics of the central nervous system. The trunk concerned the generalized species which would transmit this process of complexification of neurogenesis and followed an acceleration from the final Paleocene. And this trunk was the ancestral lineage of Man which became its apex and the current process of neural complexification.

The theory of a Paleocene simiiforme phylum has never been corroborated.

Teilhard's paleontological study nevertheless marked a major milestone in paleoprimatology, and the American George Gaylord Simpson (1902–1984), one of the great figures of 20th century in paleontology, paid tribute to him by giving his name, *Teilhardina*, to the oldest fossils of primates that lived in Asia, Europe and North America, from the Upper Paleocene to the Lower Eocene (between 56 and 47 million years ago). Today, the oldest primates are dated to 55 million years and were discovered in central China in 2003 in Hubei and in 2004 in Hunan. The fossils are kept at the Institute of Vertebrate Paleontology and Paleoanthropology of China, in Beijing. The first fossil from Hubei is an almost complete skeleton of very small size (7 cm and 20 cm with the tail) named *Archicebus achilles* (Ni *et al.* 2004). The orbits are small and the appendicular skeleton is that of an arboreal primate. The second fossil *Teilhardina asiatica* from Hunan has preserved only fragments of the skull and mandible.

As long as the chordal skeleton is not straightened and since locomotion is not the mechanical cause of straightening, these locomotor adaptations do not allow us to speak of simiiforme features. It is the simians, characterized by the straightened embryonic cord, that have inherited these arboreal characters. It would be interesting to compare the axial endoskeleton of *Archicebus* with the current species of *Loris* in Southeast Asia. The last study in 2019 confirms, using dental characteristics, the distinction between these two Chinese fossils (Morse *et al.* 2019).

Pierre Teilhard de Chardin defended his thesis entitled "*Les mammifères éocènes inférieurs français et leurs gisements*" ("French Lower Eocene mammals and their deposits"), which earned him international renown, at the Sorbonne University in April 1922. He was 41 years old and received the prize of the *Société géologique de France* (Geological Society of France) and the Roux prize of the *Académie des sciences*. He was elected President of the *Société Géologique de France* in 1926.

2.4. Human paleontology is a branch of planetology

Teilhard was in regular contact with Henri Breuil, himself trained by Albert Gaudry, and thus steeped in Lamarckian doctrine. Teilhard understood that the cosmos, which he contemplated from the volcanoes of Auvergne, was in reality the state of a disappeared cosmogenesis, just like the history of the Earth and the complexification of the life at the origin of the biosphere. The fruitful union between environment and organism did not shock him. On the contrary, for a mind that felt consubstantial with nature, like Leibniz and Goethe, it was conceivable to believe in a kind of cosmic super-organism. But the universe was no longer a being as in Haeckel's monism, and at the present moment, the universe is the stage for multiform growth. Terrestrial organisms are thus compositions built by the physico-chemical laws of their environment, and the essence of consciousness relates to these specific laws which allow the organization of complex autonomous systems. The human being, as a collective reflexive consciousness, was seen as a reflection of this growing being on the Earth, potentially able to become conscious of themselves. But the stage of growth of this cosmic consciousness reached a critical state, as the "planetarization" of conflicts testified. Theological heresy was thus consumed, the truth was no longer in the old texts written by human hands for more than 2,000 years. The authenticity of a cosmic purpose remained the only question which justified, in his eyes, all the sacrifices at the planetary scale; Teilhard set out to discover it, for the first time since the philosophers of the Antiquity, who had not conceived of the necessity to prove it.

Marcellin Boule wanted Teilhard to meet the holder of the chair of Mineralogy at the Catholic Institute of Paris, where higher education in the sciences was still poorly developed. It was important that natural sciences and evolution be taught in private establishments run by the clergy. Father Gaudefroy offered him the chair of Geology, pending the nomination of a professor; Teilhard accepted and became a lecturer. He taught geology there from 1920, defended his thesis at the Sorbonne in 1922, and was finally appointed Assistant Professor of Geology and Paleontology at the *Institut Catholique de Paris*. Teilhard de Chardin was therefore a professor and continued his teaching during the 1922/1923 academic year, while in agreement with the Society of Jesus, he devoted himself to paleontology in the basement of the *Muséum*'s gallery and participated in international conferences.

His teaching was innovative, he was not content to simply teach what he had just learned at the Sorbonne. He thought about the mechanisms of evolution on the large scales of space and time, the "continentalization", or "granitization" (silicification) of the Earth's crust, and the complexification of organisms which form the biosphere, as far as Man, who discovers himself a generic link with the physico-chemical transformations of the planetary surface. Here, we find Buffon again, but with the search for a physico-chemical, even telluric link (electric and magnetic fields) between the dynamic organization of molecules to form monocellular and then pluricellular autonomous units. The Earth's crust moves thanks to tectonics (only vertical, for the geologists at this time) and was dragged into the irreversible "drift" of its cooling toward a state of equilibrium, while climates changed and fauna migrated, but with an additional constraint: the surface of the Earth is finite, closed in on itself and the shape of the continents is, in itself, a factor of selective pressure on populations. Peninsulas are dead-ends, isolation routes, while intercontinental bridges are routes of diversity. Teilhard's evolutionary school of thought was being built: according to him, there is a planetary engine with which the increasing complexity of organisms was associated, from the tectonics of the continents and seabed to the dispersion of the oceans.

This intelligibility of an earthly planetology approached as a growing living organism is already perceptible from the earliest years of his teaching. In 1921, Teilhard published the beginnings of his evolutionary thought in an article for the *Études*[2], "How is the question of Transformism posed today?":

> Not only the fact, but the very mechanism of evolution seemed clear: to explain the metamorphoses of life, it was enough to have recourse to adaptation or natural selection, and to heredity, already a little schematized, the figure of Transformism from Lamarck to Haeckel [...]. The lines formerly drawn disintegrated, so that new explorations brought to light a plethora, in the geological layers, the remains of absolutely new animals, which forced the multiplication of families and zoological orders. Leaves began to hide the twigs, and the twigs, which were too numerous, increasingly hid the branches. For that reason, life became more and more overwhelming for the classifiers, by the richness of its forms. It soon had to be admitted that it was terribly capricious, and inordinately old in its developments. It was first necessary to give up the idea of a regular, continuous, total evolution [...]. The horseshoe crabs of the Pacific are irremediably fixed beings, which did not deviate, in a single important feature, from the type they had in the Secondary, Carboniferous or even Cambrian. This is curious. Even more disturbing, the immobilized types were not species stuck in a kind of morphological impasse. The Malaysian Tarsier or the Lemurs of Madagascar (their morphotype) could have played the role of morphological intermediaries (between the most primitive mammals and the monkeys). We now know true monkeys in the Oligocene, everything is older than we thought, in the world of life. And everything is much more stable too... [...] On groups of ungulates and carnivores, we see it beyond doubt, there are precise, simple, constant rules that preside over the gradual and "directed" complication of organisms, the precious notion of oriented variation [...]. From the smallest detail to the largest sets, our living universe – like our material universe – has a structure, and this structure can only be due to a phenomenon of growth. This is the great proof of

2 Catholic review founded by the Jesuits in 1857.

> transformism, and the measure of what this theory has definitively acquired […]. From many points of view, a radiolarian, a trilobite, a dinosaur are as differentiated, as complicated as a primate. On the other hand, their nervous system is much less perfect. Should not we look in this direction for the secret law of development? Should we not say that the main stem of the zoological tree has constantly moved in the direction of the largest brain? […] *What makes transformism is not to be a Darwinist or Lamarckist, a mechanic or vitalist, what today's naturalists hold dear is the fact of a physical connection between the living.* (Teilhard de Chardin, *Comment se poser aujourd'hui la question du Transformisme*, 1921b, author's translation)

Teilhard had always thought like a geophysicist: for him the appearance of life was a planetary, and therefore a cosmic phenomenon of geophysical origin. For him, as for Buffon (within the limits of the diversity of his "internal mold"), the history of the Earth's crust had thus shaped biological evolution, tectonics gave an account of the evolution of species on a planetary scale and he saw the spherical enclosure of the Earth as a constraining physical condition:

> The continents are natural units of the Earth's crust, so that the problems of the Biosphere can be studied there […] paying attention to the organo-plastic action exerted on animal and plant forms by the Continental Environment (inorganic or living) in which they develop. (ibid., author's translation)

The paleontologist was so convinced that he pushed for the creation of a laboratory of "Continental geology applied to the origins of Man" at the *École Pratique des Hautes Études* in 1938, housed at the *Institut de Paléontologie Humaine* (IPH) in Paris. He then created the term "geobiology" for the small Institute that he set up in 1942 in annexes of the French Embassy in Beijing.

For Teilhard, paleoanthropology was a discipline of terrestrial planetology, because the emergence of life and its complexification, whatever its rhythms, was a process of growth on an astrophysical scale which started with particles (quanta), then with atoms and, finally, molecules. Current and fossil species were thus the result of organizational plans which increased in complexity and which would not have been viable without the complexification of the nervous system to control their internal balance and their relationship with their environment. The last organizational plan with complexification of the nervous system is that of mammals. The lineage from the first primates to *Homo sapiens* was the extension of this growth channeled through the whole of the central nervous system. But Teilhard retained only the endocranial casts described by his contemporaries and did not take into consideration bipedal locomotion, because the skeleton was arbitrary, divided into cranial and postcranial territories. The bones were not seen as forming a system structured around the axial endoskeleton that protected the central nervous system carried by the appendicular skeleton. We thus understand better why the "sphenoid-cervical hinge" is unthinkable in paleoanthropology and how much the straightening of the dorsal cord constituted a huge gap in the map of the hominization process.

Thus, since Aristotle, the place of Man taken as the objective reference of a ruler has never been invalidated. The discovery of evolution has not contradicted this obvious statement, but it was not theologians who rejoiced in it, quite the contrary. The sovereignty of the naturalist had made it possible to link the gradations of the current horizontal rule, by showing that each one of them was the instant of a duration which sunk all the more deeply into the terrestrial strata the more it approached zero, or, here, the origins of life. No paleontologist has been opposed to this observation of increasing neural complexity since the first Chordates (which include Vertebrates).

2.5. China, the promise of very ancient mammalian and human species

Marcellin Boule counted on Pierre Teilhard de Chardin to succeed him to the chair of Paleontology. Emile Licent maintained his correspondence with his Jesuit colleague in 1921, urging him to come to Tien Tsin. He was an unofficial advisor to the Ministry of Agriculture and Forestry in Beijing. When Teilhard did not return his letters, he turned to his Swedish geologist colleague, Johan Andersson (1874–1960), who was attached to the laboratory of Carl Wiman (1867–1944) at Uppsala University. Andersson was also an advisor to geologist Ding Wenjiang (1887–1936), director of the Mining Bureau of the Ministry of Agriculture in Beijing, and at the head of major paleontological prospecting programs organized with the Americans at the New York Museum.

A fossiliferous karst system had just been discovered under his direction, at Zhoukoudian (Choukoutien) not far from Beijing, by two paleontologists, the Austrian Otto Zdansky (1894–1988), attached to his laboratory, and the American Walter Granger (1872–1941) of the American "Andrews" expedition. They collected fauna from the Lower Pleistocene (more than 700,000 years old) and a stone tool in quartz. Zdansky found a human-looking molar, but, refraining from talking about it, he brought it back with other fossils to be cleaned in Carl Wiman's laboratory, where he would find a second tooth in the collection. Licent persisted and wrote to Teilhard on August 13, 1922 from the Ordos, in the great loop of the Yellow River (Huang-Ho), south of the Gobi Desert. The letter was accompanied by fossils of an unknown genus of giraffe. At the same time, Teilhard was in Brussels for the 13th International Congress of Geology, during which he met Wong Wen Hao (1889–1971), a colleague of Ding Wenjiang. The two Chinese geologists had just created the Geological Survey of China. Wong spoke French and had obtained a doctorate in geology in 1912 after training at the Catholic University of Louvain (Belgium). Particularly involved in the development of the new Republic of China founded in 1912, he was Vice-President in 1947 and Prime Minister of the government of Chiang Kai-shek in 1948[3]. Wong described the research organization to Teilhard, explaining that the Geological Survey of China worked in close collaboration with paleontologists and geologists from Uppsala and the American Museum of Natural History. They were expecting good harvests of Tertiary mammals with the "Andrews" expeditions. Their programs also included the collection of old hominids from the Zhoukoudian karst. Edouard Licent was thus well integrated into the geologist-paleontologist community of Beijing with large-scale American projects. His letters sent to Marcellin Boule were the unexpected chance for France to engage in this new great exploratory phase of continental Asia where the British Empire had not extended its colonies.

When Teilhard received Licent's letter a few months later, he realized the chance for the *Muséum* to collaborate with the biologist from the Jesuit Mission in China. The Professor of geology at the Catholic Institute of Paris could become the leader of the research on the evolution of mammals in continental Asia with the *Muséum* and the IPH. On November 20, 1922, he formalized the cooperation by publishing a note on giraffes at the Academy of Sciences and convinced Marcellin Boule to organize the first "French Paleontological Mission in China" as soon as possible. In 1923, an agreement was reached between the *Muséum* and three financial backers, the Ministry of Public Instruction, the *Académie des Sciences* and the IPH. Teilhard de Chardin was officially attached to the IPH as a geologist-paleontologist with a spacious office reserved for chair professors and which was assigned to him until December 1954.

He was therefore commissioned on behalf of the IPH to enrich the collections. In accordance with the Jesuits, the single specimens would be sent to France, while duplicates would remain at the Museum of Tien-Tsin. The latter would be entirely financed by the French State (French Embassy and Mission for the fieldworks and studies of Teilhard). The reasons for his visits to China were exclusively scientific.

Teilhard left the Catholic Institute of Paris during the summer vacation and planned to resume his classes at the start of the 1923 academic year, unaware that he had just lost his status as professor. A confidential note written on the dogma of "original sin" confronted with the simian origins of Man was sent to the Holy Office which immediately suspended his teaching duties. Teilhard joined Licent in Tien-Tsin to explore the Ordos loop of the Yellow River. As early as 22 July, 1923, they discovered the first Chinese prehistoric locality *in situ* at Shui-Dong-Gou

3 Leader of the Republic of China between 1928 and 1949.

(Choei Tung K'eou). They returned with three tons of fossils loaded on 30 donkeys and mules and sent 100 kg of stone tools to the IPH. The prehistoric collection was presented to the Academy of Sciences by its perpetual secretary, Alfred Lacroix (1863–1948), geologist at the *Muséum*, who obtained the funds for the Mission. The lithic technology dated from the ancient Upper Paleolithic (40,000–25,000 years). This presentation marked the official birth of Chinese prehistory. Lacroix wrote to Teilhard to announce the continuation of the Mission on the orders of Marcellin Boule and the Jesuit Superior General. Relations with the Holy Office were tense, and he was forced to spend the winter of 1923/1924 between Tien-Tsin and Beijing.

In France, Louis Vialetton was deconstructing Lamarcko-Darwinian gradualism. Teilhard took advantage of this discovery to affirm his position and wrote a text in 1925 entitled "*L'hominisation. Introduction à une étude scientifique du Phénomène humain*" (Hominization. Introduction to a scientific study of the human phenomenon), which was unpublished until 1957:

> Illusion, the ordered, organized, inescapable distribution of the living through time and space? This I deny with all the strength of my paleontological experience. There is a natural (i.e. scientific) reason for the phenomenon of their successive appearance. (Teilhard de Chardin, *Le paradoxe transformiste*, 1925, author's translation)

Vialleton did not question evolution but rather the impotence of the Lamarcko-Darwinian model to account for the complexity of organisms, and the evolution of these organized complexities, while Teilhard was convinced of an organic uniqueness between geophysics and biophysics, between organisms and their environments considered at the biospheric and atmospheric scales of the planet. He would often tend to confuse physics with nature. Physics is a scientific discipline that studies light, phenomena of attraction (magnetization), the smallest levels of organization that always refer to forces of attraction, cohesion or repulsion, electromagnetics, waves and/or particles. This science extrapolates what the eyes do not see with equations, and, for evolution, a science confronted with the phylogenetic and irreversible self-organization processes that cannot be reproduced experimentally and that will escape measurement forever. Evolution in the past is not an illusion thanks to fossils, but its explanation will always be a theory limited by the absence of its object of study: the scientific reconstruction of a series of successive ontogenetic transformations, will always be that of a disappeared duration.

Teilhard worked over long durations, which made all the difference with regard to the hyper-specialization of paleoanthropology at the end of the 20th century and the beginning of the 21st century, which is limited to very short periods, or to very specific taxa such as the Neanderthal, *Homo erectus* or *Homo sapiens.* Deprived of geochronological hindsight, the specialization on a short lineage of taxa is thus unable to identify an emergence process involving the central nervous system and its straightening dynamics. Limiting the study of a fossil to its ecological context would result in the simplification of the interpretation of innovative traits by placing them all at the same level: that of the variety of a trait at a given instant t, without considering the structural context, which obeys neither the same rhythms nor the same morphogenetic rules influenced by past history and stored in memory in sexual cells.

The Earth is a dynamic system that evolves ineluctably toward a state of equilibrium because of the cooling of its core; for Teilhard and the geologists this reality must never be lost sight of, under the biosphere the planet is cooling and the crust moves accordingly. Hominization is part of the irreversible organic co-evolutions of this planetary geological scale, it continues the mega-evolution that generated the increasing complexity of the nervous system and of the organization that it controls, this evidence is no longer perceived today because of the hyper-specialization on fossil anatomical details.

For Teilhard, the formation of the biosphere was prolonged by the "anthroposphere". He distinguished the place of the human phenomenon in this mega-evolution thanks to long geological durations. Hominization was said to be orthogenetic, or oriented, and he called it "*orthogenèse de fond*" or "background orthogenesis", insofar as it prolonged the complexification of the central nervous system. Teilhard developed his thought by comparing the human lineage from the prosimian stem with other mammalian phylums he studied. The Paninae (gorilla and chimpanzee) and the hominins have a common ancestor, but Paninae's lineage diverged from this "background orthogenesis" 8 million years ago, forming divergent branches, whereas hominins were the manifestation of this "background orthogenesis", they were the apex of the growing trunk. And China was going to prove it with what the eyes see immediately: the organization of the cerebral hemispheres.

2.6. Peking Man: a small brain but well-cut tools

Forced to stay in China in 1923, Teilhard became closer to the scientific community that met regularly in Beijing, which was divided into foreign legations. He reconnected with Wong Wenhao, who he had met in Brussels, and met Johann Andersson's team, who organized the study of the Zhoukoudian karst. A Canadian doctor, Davidson Black (1884–1934), taught at the Peking Union Medical College, the nascent American Faculty of Medicine built in 1921 thanks to the Rockefeller funds. Black held the chair of Neurology and Embryology and later became head of the Department of Anatomy. He had been interested in the origin of man since he took anatomy courses with Elliot Smith (1871–1937) at the University of Manchester, when he described the remains of Piltdown Man. Black hypothesized the original cradle of the Tarsiiformes by visualizing the Earth from the North Pole, showing how the fossiliferous and present-day zones of Southeast Asia, North America and Western Europe converge toward Tibet. His article was published in Beijing in 1925 (Black 1925). The theory of a vast cradle of hominization in Central Asia took shape, supported by the Miocene great apes of the Himalayan foothills. Black was convinced of the existence of a hotbed of emergence in the final Tertiary, going back 1 million years, with Tibet as the epicenter, surrounded by India to the south, the Tarim Basin to the north and the Yunling or Youn-Ling mountains to the east (Yunnan). This was therefore supported by paleontological data (*Sivapithecus*, *Pithecanthropus*) and primatology, which observed a concentration of species derived from large groups of primates: Tarsiiformes, Hylobatidae (gibbon) and the great apes (orangutan). Africa did not offer anything equivalent; Charles Darwin did not write that the origins of Man could be found only in Africa, he wrote that the cradle could be in Europe, Asia and, indeed, why not in Africa, because of the presence of chimpanzees and gorillas (Darwin 1871).

China took new measures and prohibited the sending of fossils abroad. In 1926, Teilhard was definitively excluded from the chair of Geology at the Catholic Institute, and Marcellin Boule was forced to resign, since Licent was no longer allowed to send the collections. His only attachment remained to the IPH and he wanted the Institute and the *Muséum* to contribute to the development of paleontology in China. In October 1926, the Crown Prince of Sweden, Gustav VI, who was also an archeologist, was received by the Geological Survey of China. A colloquium was organized in his honor at the Peking Union Medical College to present the two discoveries which account for the founding of prehistory and perhaps human paleontology in China. Andersson was hoping for a financial commitment from Sweden for the Zhoukoudian excavations. Teilhard was exhibiting Shui-Dong-Gou, he represented the chair of Paleontology of the *Muséum* and the IPH which had a dominant position in Europe. To everyone's surprise, Andersson projected the photos of the two molars that had remained secret, giving Wiman's descriptions of them. He showed the casts to Black, immediately convinced. In the absence of excavation, Teilhard, already having been burned by Marcellin Boule's disbelief from Piltdown, remained cautious. The *Muséum* and the IPH validated neither Piltdown nor Java Man; the oldest recognized Hominidae was *Homo heidelbergensis*. Asked for an opinion on the basis of the photos alone, he answered imprudently that no one would find Man associated with a tertiary fauna, the tooth was that of a carnivore.

Since the fossils remained in China, Sweden disengaged. Black was convinced of the great paleoanthropological value of the Zhoukoudian karst, which he believed had been occupied by Hominidae 1 million years ago. He convinced Wong, Deputy Director of the Geological Survey, to entrust the paleontological study to his Department of Anatomy, maintained by the Rockefeller Foundation. Black and

Andersson informed Wiman of the agreement of the Chinese government. Thus, Black took up the torch. For his part, Teilhard informed Boule of the discovery of these two teeth and warned him, because he was far from being convinced (Teilhard de Chardin 2004).

Without delay, Black published a note in the journal *Nature* in November 1926, affirming the presence of Man in China since the end of the Tertiary era (Black 1926). He obtained funding from the Rockefeller Foundation for 1927 and 1928. Wiman remained a partner and sent a paleontologist, Birger Bohlin (1898–1990). Teilhard still had the support of Alfred Lacroix, who obtained the Loutreuil funds from the Academy of Sciences to maintain the French Mission in 1927. Davidson Black entrusted him with the study of fauna, which was no longer allowed to leave China.

Thus, thanks to the Academy of Sciences of the *Institut de France*, Teilhard ensured the first contribution of the French Mission to the study of Zhoukoudian and strengthened the links between the IPH and the Geological Survey of China via Davidson Black. The Shui-Dong-Gou collection was published in 1928 in the fourth issue of *Archives de l'Institut de Paléontologie Humaine*, entitled "*La préhistoire de la Chine*" (The Prehistory of China), signed by Marcellin Boule, Henri Breuil, Emile Licent and Pierre Teilhard de Chardin.

A new chapter of human paleontology and prehistory was being written in Northern China.

The site of Zhoukoudian was an ancient cave whose vault had collapsed, revealing a large sedimentary filling over a depth of 20 m; the site was monumental. The work was carried out under the direction of Bohlin and the stratigraphy had to be followed by the geologist Li C. Foreigners were not allowed to excavate, thus, only the Chinese were able to make discoveries. Three days before the end of the 1927 season, the diggers found a human molar in the location of the first two teeth, which were the only anthropological record. But the excavators had not yet reached the layers in place, i.e. the deposits not disturbed by erosion. Black published a detailed study and associated the name Zdansky with the new taxon, *Sinanthropus pekinensis* (Black and Zdansky) (Black 1927) by revising the stratigraphic age. Human teeth would date from the base of the Quaternary, 900,000 years at that time.

In 1928, about 30 villagers excavated 6,000 m^3 of deposits by filling more than 1,000 fossil crates with numerous human teeth, an infantile mandible fragment, a large mandible fragment with three molars and an adult parietal, from different levels. This was Davidson Black's consecration.

In 1929, the Geological Survey created the Cenozoic Research Laboratory, directed by Ding. Teilhard was trained in cave excavations, and Wong wanted to include him, negotiating his departure from Tien-Tsin with Licent. The new scientific team was composed of the paleontologist Young Chung-Chien (1897–1977), nicknamed "the father of Chinese paleontology", a young Chinese geologist, Pei Wenzhong, who would become the "father of human paleontology" in China, Davidson Black, honorary director of the laboratory and in charge of the study of Hominids, and finally Pierre Teilhard de Chardin, appointed advisor and supervisor of the excavations, in charge of the study of all vertebrates, excluding Hominids.

Pei Wenzhong took part in the excavations: he went down the well to a depth of more than 30 m and himself dug with the villagers in the coldest weather. On December 2, 1929, he extracted the first almost complete neurocranium from its mud gangue. Then, a second incomplete skull was discovered on December 28. Davidson Black estimated the number of Hominids to be 10 individuals, including an adolescent skull and a reconstructed adult skull, mandibles, numerous teeth, fragments of ribs and the skull of a child in a very hard breach. Teilhard expected the skeleton to be complete. In any case, this was the oldest known infantile stage of a hominid. These fossils came from different layers: the sub-adult skull, or *Sinanthropus* III of Locus E, was the oldest. Recent dating gives an age of 770 ± 0.08 ka (Shen *et al.* 2009). *Homo heidelbergensis* was dethroned. In 1929, this skull became the oldest hominid in continental Eurasia.

Davidson Black carried out an exhaustive study of the skull. This one was exceptional because the sutures were not synostosed, he was thus able to disarticulate the bones and give a detailed description of their internal face. He took X-rays and made casts of each of the bones, as well as a plaster cast of the impressions of the brain (Black 1931). The monograph was remarkably illustrated by photographs of excellent quality with a tracing drawing the contours and with the names of the different anatomical structures made for each plate. The central part of the base was missing, so it was not possible to measure the sphenoidal angle, but the high position of the cerebellar fossa and the inclination of the cerebellar face of the petrous pyramids gave solid indications. The skull was at an advanced stage of growth because the supraorbital torus was already well formed. The cranial capacity was 964 ml; Black attributed this to an adolescent (Black 1933). This skull may well have been that of a young female in the late stages of growth. The IPH acquired these casts in 1932. Only three original series existed in the world, the other two being kept at the Museum of Natural History in New York and at University College, London. However, these casts are unique because all of the fossils disappeared in 1941 when it was decided to protect them from Japanese looting during Second World War.

Black and Teilhard discussed the cognitive abilities of the *Sinanthropus* on the basis of brain imprints and meningeal vascularization. The French paleontologist was doubtful, due to the lack of lithic tools. Intrigued, he questioned Pei. The latter had indeed put aside a quartz stone which turned out to be cut. Teilhard decided to excavate another cavity (Kötzetang or Gezitang). The results were immediate: from 1931 to 1932, the excavators collected many quartz tools, coal and new fossils of *Sinanthropus*, including mandible, teeth and clavicle. As early as 1931, Black invited Henri Breuil to come and study the operation. In return, Pei Wen Zhong went to Paris to register for a thesis in order to learn the study of lithic industries at the IPH, which he defended at the Sorbonne in 1937. These were the oldest prehistoric lithic tools. For the first time, the organization of cerebral hemispheres more complex than a great ape, likely to illustrate cerebral hominization, could be confronted with the materialization of gestures following abstract reasoning. At first, Teilhard and Boule disagreed, because the latter did not conceive that a 1,050 ml brain could be capable of such a manual and conceptual feat.

A tragedy occurred on March 15, 1934: Davidson Black died in his laboratory, struck down by a heart attack. Teilhard now had to take over the direction of the programs. The chair of Anatomy at the Peking Union Medical College was vacant, and it was the German physician and paleoanthropologist Franz Weidenreich (1873–1948) who took over in 1935. He became well known in 1928 through the study of a Neanderthal neurocranium discovered in the paleoanthropological site of Weimar-Ehringsdorf, known since 1914 (Weidenreich 1928). He was interested in *Sinanthropus* and was willing to leave the chair of Anthropology at the University of Frankfurt. Excavations continued and nine incomplete human skulls were collected between 1936 and 1937.

2.7. The first study of the internal basis of a Hominidae fossil

In Java, the skulls of *Pithecanthropus* also started to accumulate. The German anthropologist Ralph von Koenigswald (1902–1982) participated in the excavations of the Geological Survey of the Dutch East Indies from 1931 to 1933, which focused on an ancient terrace of the Solo River at Ngandong. The collection included 11 skulls. In 1939, Weidenreich and Koenigswald met in Beijing to discuss the comparative anatomy of the fossils that constituted the largest collection of East Asian hominids, consisting of 17 almost complete craniums, 6 in Beijing and 11 in Java. The skulls presented a morphological unit that made it possible to consider them as being part of the same genus and species. Koenigwald suggested to keep the first name, *Pithecanthropus*. But no skull had preserved its base intact (Weidenreich 1937). No sphenoidal angle could be measured, and attention was focused on the most visible of the morphological evolutions. As is often the case, this was the most misleading link in the chain of causalities. These are obviously the cerebral hemispheres, or the hypertelic (excessive) development of the embryonic telencephalon. From this point on, anthropologists would consider telencephalization. As the skull base and the post-cranial skeleton were almost absent, the search for correlations with the spinal column was not contemplated. The sphenoid and the brain stem were left out, and embryogenesis was totally ignored. The study of hominization focused on the brain and adult technical intelligence without considering a physiological relationship with the cerebellum. The latter was reduced to the less noble functions of the unconscious control of postural balance.

Franz Weidenreich undertook a systematic study of the skulls of *Sinanthropus* by looking at the evolution of their base, comparing them with *Homo sapiens* and chimpanzees (Weidenreich 1943). The sphenoidal angle could not be measured on the fossils, since the anterior and middle brain fossae were missing in all the specimens. The angle was open in chimpanzees and closed in *Homo sapiens.* He suggested a geometrical reconstruction by referring to the position of several homologous cranial landmarks. On the external surface, he took the most anterior edge of the occipital great hole (basion) and the upper margin of the external auditory meatus (porion). On the internal surface, he took the internal auditory meatus, an aperture visible on the voluminous petrous pyramid that formed the anterior wall of the cerebellar fossa (Figure 2.3). What was the degree of straightening, or verticality, of the brain stem? Weidenreich had compared the position of these landmarks by orienting the skull on a line joining the root of the nose (nasion) to the furthest margin of the great occipital hole (opisthion). On *Sinanthropus*, the inner and external meatus are slightly above this line, whereas in *Homo sapiens*, they are much higher, with the anterior margin of the great occipital hole further forward.

He called this difference “deflection” and concluded that it was clearly more pronounced in *Sinanthropus* than in Paninae, but its sphenoidal angle should be more open than in *Homo sapiens* (Figure 2.2(A)). This difference between the late stages of growth in *Sinanthropus* and *Homo sapiens* corresponds to a lesser verticalization of the clivus and cerebellar surface of the pyramids. Then, by comparing adult cranial profiles of the gorilla, *Sinanthropus* and *Homo sapiens*, he highlighted a transverse axis of rotation passing through the center of the base of the skull (Figure 2.2(B)), just in front of the external auditory meatus. The face and neurocranium seem to roll up around this axis, the circles narrowing and all tissues contracting in its axial center. Later, at the American Museum of Natural History in New York, which he joined in 1941, he compared the degree of inclination of the pyramids of a skull of Ngandong (Java) with the gorilla and *Homo sapiens*, and found a cerebellar face of the pyramids more upright, but less vertical as on the *Sinanthropus* (Weidenreich 1951).

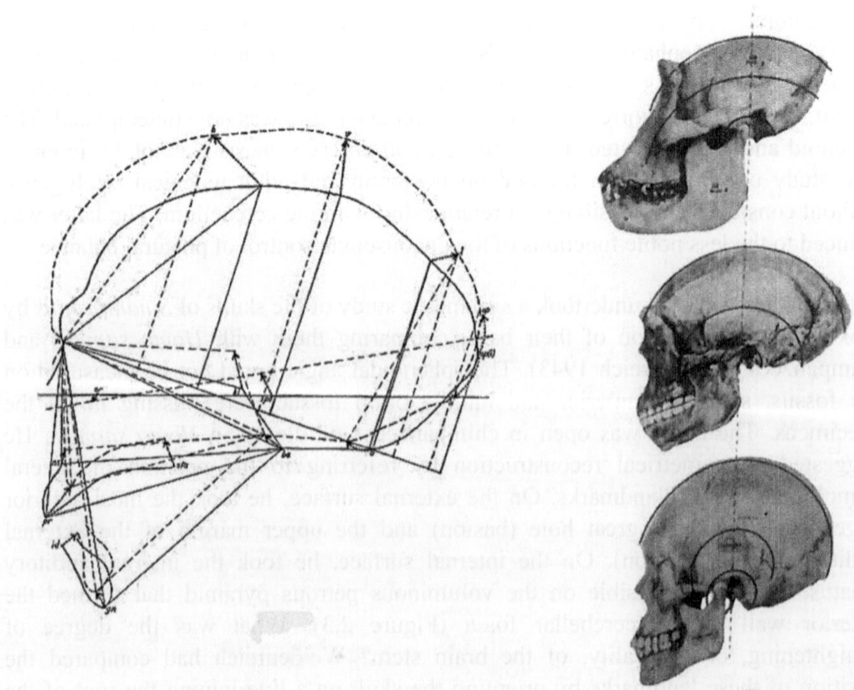

Figure 2.2. *A: Diagram comparing the sagittal profile of Sinanthropus III and Sapiens (b: bregma, ba: basion, i: inion, kl: klition, l: lambda, op: opistocranion, or: infra-orbital, po: porion) (Weidenreich 1943, plate LXVIII). B: The rolling-up of the skull around a transverse axis of rotation at skull base compared between the gorilla, the* Sinanthropus *and* Homo sapiens *(Weidenreich 1939) (photo credit: Library of* Institut de Paléontologie Humaine*). For a color version of this figure, see www.iste.co.uk/ dambricourt/embryogeny.zip*

Weidenreich interpreted this "deflection" by considering the general shape of the vault. The brain of *Homo sapiens* is more coiled in its occipital part, so he supposed that the increase of the cerebral hemispheres would have pushed the cerebellum, and thus the cerebellar fossa with its great occipital hole, forward and downward. Nothing is said about the sphenoid.

2.8. A new paradigm: telencephalization

The degree of straightening of the pyramids and the clivus, which form the anterior wall of the lower cranial fossa, would simply be the consequence of a development of a brain and cerebellum that became more complex very gradually, passing from a hypothetical archaic stage *Homo sapiens* to modern *Homo sapiens*, or Cro-Magnon Man (Figure 2.3). Since this first comparison, paleoanthropologists and anatomists interested in the neurocranium of fossilized Hominids have taken this speculation for granted, as Grover (1962) reminds us. Telencephalization became the new paradigm of hominization.

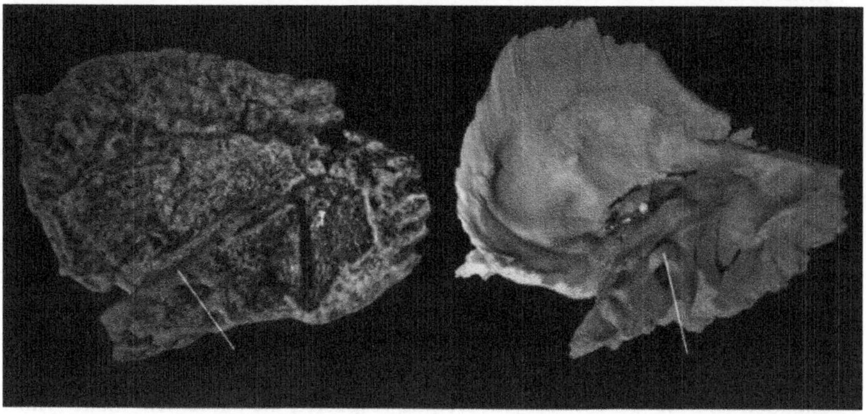

Figure 2.3. *Comparison of the straightening of the right petrous pyramid between* Sinanthropus III (Homo erectus pekinensis) *and* Homo sapiens *(white line on the internal auditory meatus) (photo: A. Dambricourt Malassé). For a color version of this figure, see www.iste.co.uk/dambricourt/embryogeny.zip*

However, the comparison of Neanderthal skulls that were sufficiently well preserved at different stages of growth already invalidated this speculation, because their encephalization with a cranial capacity higher than *Homo sapiens* did not change the inclination of the pyramids. The cerebellum did not develop forward and downward, but backward, forming the famous "occipital bun" (see section 7.4). In other words, the sphenoid, the clivus and the pyramids did not change their degree of verticality compared to the 800,000-year-old *Sinanthropus*, while its brain was clearly less developed. With neandertalization, the brain and cerebellum extended their territory at the vault level, since the ossification of membranes took place from the lobes. On the other hand, the center of the base of the skull has no contact with the cerebral hemispheres and forms a block that precedes the telencephalization in the chronology of cerebral development. Its morphogenesis preceded the formation of the hemispheres.

To realize this, it would have been enough to look at the work of embryologists at the beginning of the 20th century who had already demonstrated this. For a reason that the history of science will certainly explain, embryologists and zoologists no longer cross-referenced their observations, and paleoanthropology had never heard of the work on the embryogenesis of the primordial cranium of Man, or the chondrocranium, published in 1900 by Giuseppe Levi (1872–1965) (Levi 1900). Levi had studied medicine and surgery at the Institute of Higher Studies of the University of Florence in 1889. In 1899, he pursued research at the University of Berlin with the embryologist Oscar Hertwig (1849–1922), a former medical student of Ernest Haeckel. Three former researchers from his laboratory at the University of Turin were awarded the Nobel Prize in Physiology or Medicine, including Rita Levi-Montalcini (1909–2012) in 1986 for the discovery of the Nerve Growth Factors. We will come back later to Levi's work which is of primary importance.

Excavations at Zhoukoudian were suspended in 1937 following the conflict between Japan and China. In 1939, the United Kingdom, France, Australia and New Zealand declared war on Germany. In 1941, fearing the escalation in the Pacific between the Japanese army, allied to Germany, and the United States, American nationals had to leave China. The director of the Peking Union Medical College asked Franz Weidenreich, who was Jewish, to travel to the United States to stay there for good. Weidenreich had planned to retreat to the American Museum of Natural History in New York, where he had stayed in 1940. The Rockefeller Foundation and the US Embassy in Beijing respected the agreement made with the Chinese: human fossils had to remain on the territory.

Teilhard was opposed to the idea of sending the collections to the United States. Weidenreich feared losing them. He convinced Wong to send the fossils to New York, where they would be safe, and Wong managed to persuade Chiang Kai-shek. The agreement was made with Colonel William Ashurst (1893–1952), commander of the detachment of the North China Marine Corps and of the guard unit at the US Embassy in Beijing. Weidenreich returned to New York with the original casts, photos and drawings, which are now preserved in the American Museum of Natural History, and sent his instructions to his deputy, Claire Taschdjian, who had remained in Beijing with Teilhard. They were carried out by two Chinese laboratory technicians (Lenz 1998). Two crates were prepared, and then transported to the port of Tien-Tsin by train. The day after the attack on Pearl Harbor, on December 7, 1941, the Japanese army ordered all American property to be seized and captured marines and US Navy personnel. The crates were never recovered despite immediate Japanese and Chinese investigations. Since then, various hypotheses and contradictory testimonies have fueled hopes of finding them.

In 1942, the French Embassy in Beijing accepted the creation of the Geobiology Laboratory. Teilhard de Chardin had part of the Licent Museum of Tien-Tsin, which had been financed by French government funds, transferred to this small Institute, thus allowing the study of fossils and their publications in his new journal *Geobiologia* printed in Beijing. Japan surrendered on August 17, 1945, after the atomic bombings of Hiroshima and Nagasaki. Teilhard awaited instructions. The collections of his small institute joined those of the Cenozoic Laboratory of the Peking Union Medical College, which resumed its activities. They would be used for the progressive creation of the Institute of Vertebrate Paleontology and Paleoanthropology (IVPP). The embassy informed Teilhard in March 1946 that he could finally prepare to leave Beijing from Shanghai on the next flight to England. He returned to France determined to deepen his reflection on the "human phenomenon", after having survived the hell of the 1914–1918 trench warfare and witnessed the power of technical intelligence with the mastery of atomic energy. From the oldest stone tools to the nuclear bomb, the stakes of the amplitude of cosmogenic phenomena are woven into the Earth's crust throughout the hominization of consciousness, or a globalizing telencephalization which he called the "noosphere". According to him, the 20th century appeared to be the culmination of a meta-physical reality perceptible through cosmo-bio-anthropogenesis in a struggle against annihilation.

Marcellin Boule died on July 4, 1942. The IPH was directed by the paleoanthropologist Henri-Victor Vallois (1889–1981), who inherited both the chair from Armand de Quatrefages and the direction of the *Musée de l'homme* built in 1937 in a wing of the *Palais de Chaillot*, facing the Eiffel Tower. The succession of the chair of Paleontology caused a stir because Boule's objective was to pass on his position to a former student of Teilhard de Chardin, the paleontologist Jean Piveteau (1899–1991). The professors of the *Muséum* voted for Camille Arambourg (1885–1969), an Africanist field paleontologist (Hadjouis 2015). The African continent was becoming the new El Dorado for prehistorians and paleontologists in search of the oldest Hominidae. After 5 years of absence and following the death of Marcellin Boule, Teilhard de Chardin's spacious office at the IPH, or the "Laboratory of continental geology applied to the origins of Man", was "liquidated", to use his term, by Raymond Vaufrey (1890–1967), a geologist who had been attached to the IPH since 1930 (letter to Henri Breuil in January 1955 from New York). The cradle of Man's origins had migrated to Africa; Teilhard would never again return to Asia. South Africa awaited him.

3

South and East Africa: The New Cradle

3.1. Gracile, robust Australopithecines and Humans

A fragment of a human mandibular symphysis with a modern appearance was collected at Kanam in western Kenya by Louis Leakey in 1932, during surveys of the Miocene (6 Ma) and Ancient Pliocene (4.5 Ma) formations. He named it *Homo kanamensis*, and announced that this region of the East African Rift System was the cradle of the emergence of *Homo sapiens*, and dated back as early as the Pliocene. Teilhard considered this announcement to be too premature. The controversy prompted geologists to verify the stratigraphic context, which invalidated the geological age. Leakey maintained his position, and new investigations would show that the mandible fragment had come from colluvium. In 1939, a maxilla fragment was collected in Tanzania by a German anthropologist, Ludwig Kohl-Larsen (1884–1969), during surveys at Garusi in the Laetoli region. This fragment was identified in 1950 as that of a Hominidae and named *Meganthropus africanus* Weinert, arousing the disapproval of two paleontologists, Raymond Dart (1893–1988) and Robert Broom (1866–1951), who were able to recognize the concave sub-nasal prognathism and dentition of the *Australopithecus* known in South Africa. This fossil was a Paranthrope and not a Man.

Henri Breuil had been studying South African lithic industries in Johannesburg since 1942. He planned to collaborate with Teilhard in future studies of prehistoric sites and *Australopithecus*-bearing fossiliferous breccias. These sites studied by Raymond Dart, Robert Broom and Clarence Van Riet Low (1894–1956) acquired a certain notoriety, notably Taung, Sterkfontein, Kromdraai, Swartkrans and Makapansgat. The Prime Minister of the South African Union, General Jan Smuts (1870–1950), agreed to allow Teilhard to advise on the fauna collection and the paleoanthropological potential. His aim was to obtain funds from the Wenner-Gren Foundation (WGF) in New York and to help develop the very first research in prehistory and paleontology in the country, as quarry workers were collecting fossils and there was a shortage of experts to analyze them. The organization of systematic excavations following the stratigraphic method, as in Choukoutien, was an expensive process, but Breuil was well aware of the similarity with China with tons of sediment that needed to be extracted.

Anthropology was a recent science in Southern Africa. The University of the Witwatersrand in Johannesburg had established a new Department of Anatomy in 1922. The head of the department was Raymond Dart, an Australian anthropologist. In 1924, a quarry worker brought him a small skull still caught in a breccia. It took weeks of patience in the laboratory to get it out, and what he discovered was exceptional. The fossil was the cranium of a young anthropomorph with the mandible and the natural endocast of the brain. The fossil resembled a very young chimpanzee but with differences over the whole head that brought it closer to Man. This specimen was the first natural replica of the brain and cerebellum surface of an extinct anthropomorph (i.e. Pongidae and Hominidae). Dart described the fossil in 1925 in the journal *Nature*, and named it *Australopithecus africanus*, considering it to be a Hominidae and not a Pongidae because of the position of the occipital hole, which was lower and more advanced. British paleoanthropologists, such as Arthur Smith Woodward, did not recognize this interpretation and asserted that it could only be a young, deformed chimpanzee, as no great ape had ever lived in such southern latitudes.

In 1936, a physician of Scottish origin, Robert Broom (1866–1951), who had moved to South Africa to pursue paleontological research, was convinced of the analysis. He obtained the post of curator of the Transvaal Museum in Pretoria and became interested in the quarries being exploited in the Pretoria region, particularly in a karstic plateau at Sterkfontein that was rich in fossiliferous breccia. He asked to be informed as soon as a fossil resembling a monkey or a human appeared during the extractions. Shortly afterward, a fragment of an adult skull was extracted and this confirmed Dart's conclusions. Robert Broom extended the investigations in the region, for example at Kromdraai, where more robust specimens were collected in 1937 (Figure 3.1).

He created a new genus of *Hominidae* and named it *Paranthropus robustus*. The quarry workers at Sterkfontein continued to extract fossils, and in 1947 an intact adult skull – but one without a mandible – was excavated from the breccias, as well as parts of a locomotor skeleton, which consisted of half a pelvis and the corresponding end of the femur which articulated with its glenoid cavity. Broom and his assistant John Robinson (1923–2001) named it *Plesianthropus transvaalensis*, popularly known as Mrs. Ples. In 1948, a third fossiliferous karst with a robust mandible was discovered at Swartkrans, which he named *Paranthropus crassidens*, and in 1949 a gracile mandible was collected in the upper breccia, but this was confirmed to be a primitive human and was named *Telanthropus capensis*.

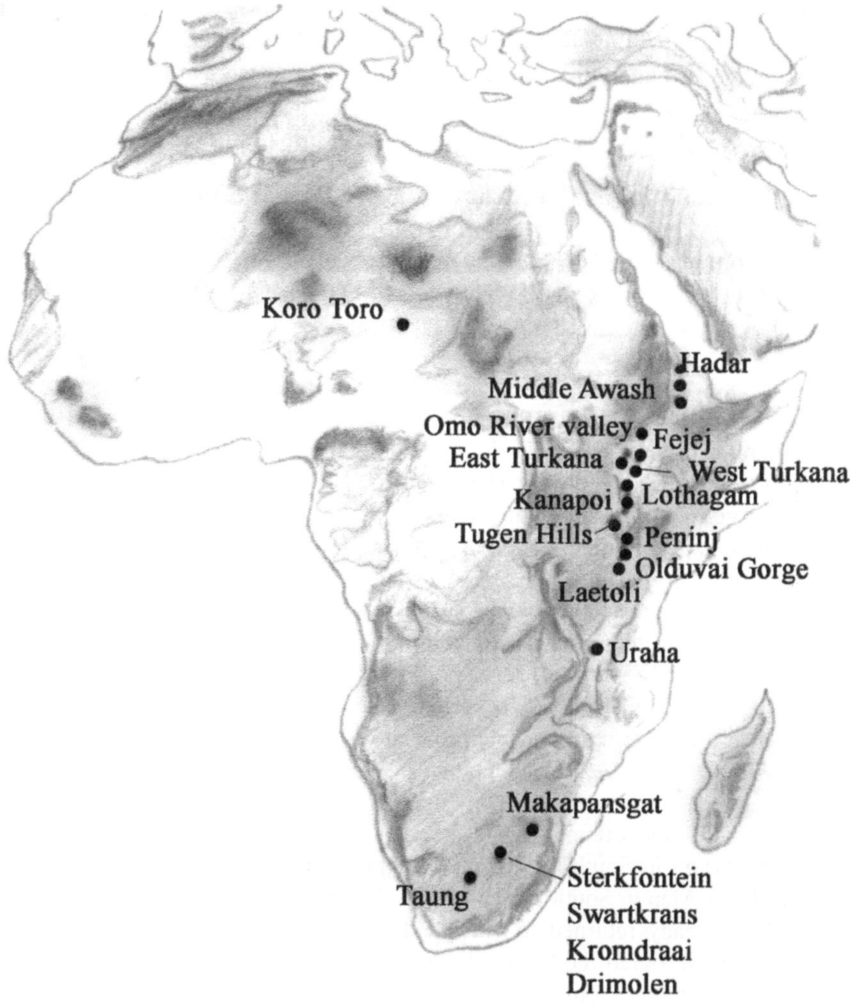

Figure 3.1. *Location of fossil hominins*

Plesianthropus was actually the adult stage of the Taung Child, an *Australopithecus africanus*. There was no longer any doubt, the internal base of the skull was lowered forward and downward, the clivus was straightened, moved downward, the pelvis had the characteristics of a Man, short and wide, it carried the mass of the body verticalized and the head of the femur was higher than the greater trochanter. The hip joint differed from any great ape. The skeleton was too incomplete to know if the body was adapted to an arboreal lifestyle, but in any case, the anatomy was neither that of a great ape nor that of Man, but like Man, the axial skeleton was verticalized.

3.2. "A systematic research plan for Early Man in South Africa"

On returning to Paris in the summer of 1946, Teilhard was already thinking of organizing an international symposium in South Africa on "Australopithecus-bearing fissures". Appointed director of research at the *Centre national de la Recherche scientifique* (CNRS, French national center for scientific research), he was about to join Breuil in July 1947 when a heart attack hospitalized him for several weeks, and forced him to convalesce for several months. The Administrator of the *Collège de France*, Edmond Faral (1925–1954), proposed that he apply for the chair held by Henri Breuil, which was officially acquired by a majority vote of the Assembly of Professors. Teilhard went to the Vatican at the end of 1948 to obtain the authorization of his hierarchy as well as for the imprimatur for his first book "*Le phénomène humain*", which had been awaiting an answer since 1940. His teaching post should have lasted 8 years, however, isolated from the field, the prospect of teaching prehistoric culture did not interest him. Only the reflection on what he called "Anthropogenesis", of which human beings were still barely aware, was his motivation. His superiors asked him not to apply. The imprimatur was also refused. Teilhard was elected non-resident member of the *Académie des sciences* in 1950. In 1951, Paul Fejos (1897–1963), director of the Wenner-Gren Anthropological Foundation, New York, and of the Peabody Museum of Archaeology and Ethnology at Harvard University, granted him a budget for a 3-month mission in South Africa. His program was entitled "Systematic Research Plan for Early Man in South Africa". When he first visited southern Africa in 1951, Breuil had already returned to Paris and Robert Broom had died a few months earlier. Teilhard toured the karst fillings with Van Riet Low and John Robinson in Taung, Sterkfontein, Kromdraai, Swartkrans and Hopefield, as well as the "Cave of Hearths" in Makapansgat. His mission completed, he went to New York to make his report, and Fejos offered him the position of advisor to the WGF, putting an office at his disposal. For his research, he was based at the American Natural History Museum of New York. In 1952, he was invited to present his analyses and conclusions to the New York Academy of Sciences and at the first International Symposium on Anthropology of the WGF. He returned to South Africa in 1953 and

1954 to support the financing of several explorations and excavations of sites that seemed promising to him, and was particularly interested in the participation of the *Muséum* chair of Paleontology, directed by Camille Arambourg, in this new phase of exploration of the origins of humankind.

Teilhard traveled throughout China, but he also prospected the Indian subcontinent in the Potwar plateau and Himalayan foothills, finding himself a *Sivapithecus* mandible, as well as central India in the Narmada Valley. He visited Javanese sites and prospected the Irrawaddy terraces in Burma. No other Cenozoic paleontologist knew South and East Asia better than Teilhard. His opinion was therefore authoritative. Teilhard de Chardin concluded that South Asia no longer appeared to be the cradle of the emergence of Man, the new Pliocene Hominids of South Equatorial Africa, which were lacking in Asia, spoke in favor of an African cradle of humankind. Teilhard had big ideas about sub-Saharan Africa. He bet on the great fissures of Makapansgat, which offered a consequent filling with *Australopithecus* at its base, then, by going up after several sterile layers, sharpened lithic tools, or hand axes, on both sides. This famous lithic industry is called Acheulean since its discovery at Saint-Acheul in the terraces of the Somme River by Albert Gaudry in 1842.

Teilhard advised a Portuguese geologist, Fernando Mouta, to excavate a fossiliferous karst in southern Angola in order to search for *Australopithecus* with Desmond Clark (1916–2002), the curator of the Livingstone Memorial Museum, and Camille Arambourg, Boule's successor at the *Muséum*. The Livingstone Museum was close to the Victoria Falls on the border between Northern Rhodesia (now called Zambia) and Southern Rhodesia (now called Zimbabwe). Teilhard obtained government approval and WGF funding for these two projects in 1952 (Arambourg and Mouta 1952). In 1953, Desmond Clark and Teilhard traveled to Northern Rhodesia (Zambia), in the Lusaka region, to check for pebble tools on an ancient terrace of the Kafue River. This would be the oldest known lithic assemblage (or pebble industry), and the Kafuan (as they became known) pebble tools were blunt and the distinction of intentional removal was not obvious. Teilhard insisted that the WGF financed Desmond Clark's excavations and prospecting for 1954. He partnered with a geologist and paleontologist from the British Museum, Kenneth Oakley (1911–1981), and obtained the funds. The WGF worked with Teilhard to organize the 1955 International Symposium on "Man's Role in Changing the Face of the Earth". In January 1955, Van Riet Low and Robinson announced to him the great news that they had found pebble industry at Makapansgat, just above the *Australopithecus*-bearing layer; "we're getting warm," he wrote to Henri Breuil. Teilhard had never felt so close to the oldest trace of human presence on the Earth. He died of a final heart attack at the age of 74 years, on April 10, 1955 in New York.

The picture that emerged in South Africa showed a variety of Hominids, that is, primates in which the segmented axial skeleton – from the sphenoid to the sacrum – was vertical. This variety was already widespread in southern equatorial Africa, while the pebble industry confirmed a very ancient human presence in Makapansgat. The issue was dating these tools and the discovery of human fossils, because the clue to a technical manual intelligence still remained a volumetric ratio between the brain and the face. Encephalization supposed a cerebral Rubicon or the crossing of a threshold which could only be arbitrary if we had to rely on the curved shape of the vault which grows linearly, without angular "fracture". On the other hand, the internal base also presented reliefs with angular crests separating the three stepped cerebral fossae and objective angular "ruptures" with verticalization. However, the verticality of the nervous system was not considered as an indicator of cognitive abilities, as it was seen as a mechanical consequence of vertebral straightening. For Teilhard, the *Australopithecus* presented a face that was too simian (prognathic) and a cranial capacity that was too small to have had such elaborate cognitive capacities. His synthesis can be summed up in this way, and it has lost none of its modernity:

> The recent discovery of Australopithecidea sheds considerable light on the regions where the human zoological form must have been formed. By their large size, by their dental characteristics (reduction of the canines, shapes of the molars), by the shape of their mandible, by their partially upright position, these simians certainly fill an important morphological void between the Anthropoids and the Hominins. Is it, as Broom says, a direct evolutionary line whose rings would all be found *in situ* in the caves of the South? We do not think so. Dr. Broom thinks he can demonstrate that this establishes a continuous zoological link between the anthropoid Ape and Hominins – the design suggesting, in his opinion, the idea of a directed plan, much more than the influence of Darwinian-type natural selection. A rather curious intellectual position in which a most radical scientific evolutionist allies himself with undisguised traces of "creationism" "metaphysical, if not even religious". (Teilhard de Chardin, *Les Australopithèques et le chaînon manquant ou "missing link" de l'évolution*, 1950a, author's translation)

Teilhard reasoned for Hominids as for any other lineages he had studied in the faunas of the Paleocene, Miocene and Pliocene. For him, to find already diversified Australopithecines was to be well downstream of the diverticulum of the first Hominids, as these were specialized divergences that had been in equilibrium with their environment for a long time. For Teilhard, Man was the least specialized of the primates (e.g. in his masticatory apparatus), which is not the case of the collected Australopithecines, which were already too late, in any case, to constitute the ancestral lineage. The archaic human form was thus already on the point of emerging, if not already emerged, and therefore it was necessary to postpone the emergence of the first Men well upstream of the radiation of the Australopithecines of the final Pliocene, that is at least 3 million years ago, and not necessarily look for it in South Africa:

> Nothing proves that the *Telanthropus* (*Homo*) is an element of the series, it could be entirely independent, in its evolution, of the Australopithecus encountered in the older breccias. On the other hand, however progressive Australopithecidea may be in their anatomy, even in their advanced type (Plesianthropus, Paranthropus), their characters remain "simian" much more than "human". The Australopithecus would represent, at the end of the Tertiary Era, for South Africa, not the main stem, but a marginal branch, nothing seems to appear to us in the form of linear series, but only in the manner of nested segments, more or less divergent and relaying between them. (ibid., author's translation)

If Australopithecines were automatically classified in the human lineage against the classical processes of radiation, this choice risked being seen as the presupposition of a finalized plan and the search for the missing link which did not exist in the fossil series:

> Dr. Broom's position will appear to many of his geological colleagues to be far too simplistic. And the same can be said of his idea of a "Plan" for Evolution. A plan … A vague expression, just as sketchy as his condemnation of contemporary neo-Darwinism. Because, finally, no matter how Lamarckist or Bergsonian one may be, one must admit that Life advances only by groping among the effects of large numbers and the game of chance. (ibid., author's translation)

This granted him the objectivity of an obvious statement, for which it was advisable not to seek an explanation in theological presuppositions, by keeping away from providential transcendental considerations. Teilhard did not distinguish between matter, already known for its wave-particle duality, and spirit, he added the internal arrow of time in the direction of growth, the thermodynamic arrow reversing the tendency toward equilibrium and death. This statement read as follows:

> The existence of a drift of organized matter towards ever more complicated arrangements, and ones correlatively more charged with consciousness, (in which) Man represents, in this particular direction and in the field of our graspable experience, the most advanced state, the end, at least provisionally, the ultimate point in an evolution

> which would not be understood outside a preferential march of the Universe towards higher and higher cerebro-psychic levels. If this is what Dr. Broom meant when he spoke of an anthropocentric orientation of Life, I imagine that tomorrow all of Science will agree with him. The future will decide. (Teilhard de Chardin, *Les Australopithèques et le chaînon manquant ou "missing link" de l'évolution*, 1950a, author's translation)

3.3. The Princeton synthetic theory (neo-Darwinism) and the Sorbonne replica (neo-Lamarckism)

Paleontologists followed the progress of genetics with the announcement of the synthetic theory of evolution formulated in January 1947 at Princeton University. But in France nothing was a foregone conclusion. Marcellin Boule had hoped that Jean Piveteau, his former student, would succeed him in the chair of Paleontology at the *Muséum* in 1936 rather than Camille Arambourg. At that time the chairs of the *Muséum* managed the collections and exhibitions, but the courses were public, and, like the *Collège de France*, the chairs still did not provide a university education, as had been the case since the creation of the Royal Medicinal Garden of Plants. Jean Piveteau was appointed lecturer in Paleontology at the Sorbonne University in 1938, then professor of Paleontology in 1942, but without a chair. He created the Sorbonne's first chair of Vertebrate and Human Paleontology in 1953.

In reaction to the Princeton symposium, Jean Piveteau organized an international symposium in April 1947 entitled "Confrontation of the points of view of the Paleontologist and the Geneticist in the face of the transformist theories", with the CNRS, the Rockefeller Foundation, Camille Arambourg, who represented the chair of Paleontology of the *Muséum*, and Pierre-Paul Grassé (1895–1985), holder of the chair of Zoology and Evolution of Organized Beings at the Sorbonne. The colloquium would take place at the *Institut de Paléontologie Humaine*, directed by the paleoanthropologist Henri Vallois. Prominent personalities were invited to share their knowledge and thoughts, including George Gaylord Simpson, David Watson, Henri Vallois, Conrad Waddington and Teilhard de Chardin for paleontology, and Lucien Cuénot, the founder of genetics in France, Maurice Caullery, John Haldane and Marcel Prenant for evolutionary biology. As an introduction to the proceedings, Jean Piveteau wrote:

> The divergence becomes clear when geneticists approach the question of evolution like de Vries' mutationism, where transformations take place by sudden saults, spontaneous in a way, without a definable direction (micro-evolution). Belgian paleontologist Dollo: evolution is discontinuous, directed and irreversible (macro-evolution). What the

> paleobiologist observes and wants to explain are these great evolutionary changes which are not made by simple quantitative modifications but by qualitative transformations, leading to true organic revisions. In passing from the small divisions to the major divisions of classification, is there not a necessary appeal to new explanatory factors? One of the problems that it would be essential to expose in a precise manner is that of the relationship between micro-evolution and macro-evolution. (Piveteau, *Paléontologie, génétique et transformisme*, 1950b, author's translation)

Parasitologist Marcel Prenant (1893–1983), holder of the chair of Zoology at the Sorbonne, described the Lamarckian school as "not very rational", giving the environment the role of a kind of designer:

> Apart from well-understood neo-Darwinism, which does not offer any difficulty in principle, any other proposed explanation of evolution goes beyond science, to fall more or less in metaphysics. (Prenant, *Les mécanismes de l'évolution*, Commentaries, 1950, p. 223, author's translation)

For paleontologist David Watson (1886–1973), professor at University College London, the distinction between the different major organizational plans did not require any more knowledge:

> I do not think that the origin of large groups, macroevolution, calls for causes other than those already successfully applied to more restricted orders of facts. (Watson, *Paléontologie, génétique et transformisme*, Commentaries, 1950, p. 232, author's translation)

The British geneticist John Haldane (1892–1964) analyzed the different positions and cases:

> A paleontologist is not linked to any one theory. He can be a Lamarckist or a Darwinist. He can postulate a theological intervention, like Wallace and Broom. He can be wary of any explanation of the mechanism of evolution as still premature. In the field of genetics, I believe that all those who have proven this admit the Mendelian heredity of at least some traits [...]. Extranuclear heredity exists, especially in green plants [...]. The Heir has proved that the only extranuclear hereditary trait discovered in *Drosophila melanogaster* can be transmitted by grafting, and its agent can be considered as a virus. (Haldane, *Paléontologie, génétique et transformisme*, Commentaries, 1950, pp. 184–185, author's translation)

The American paleontologist George Gaylord Simpson (1902–1984) from the *American Museum of Natural History*, who thought himself a great specialist of the statistical calculations newly introduced in genetics and orthogenesis in paleontology, stated:

> The reconstitution of geologically long orientations in a single direction is subjective […]. The postulates inherent in one form or another of the orthogenetic theory […] by unknown factors are neither required nor probable according to the present data […]. The driving role of natural selection, in this synthetic view, is not limited to a channeling, without a restrictive meaning, but it is also creative, by giving rise to a progression in a defined direction, orthogenesis, as it actually occurs, is orthoselection. (Simpson, *L'orthogenèse et la théorie synthétique de l'évolution*, in *Paléontologie, génétique et transformisme*, 1950, author's translation)

An example of the evolutionary trend was the increasing neocorticalization in mammals, well described in horses over a 55-million-year period by Tilly Edinger (1948). Simpson was contradicted by paleontologists familiar with orthogenesis that could not be explained by the statistics of random mutations and retained only because they would have presented a totally accidental advantage. *This is the basis of the greatest of the polemics that divided the neo-Darwinists of Princeton, who thought they had an unsurpassable explanation (evolution is exclusively genetic and by chance), and the palaeontologists in search of processes far more complex and that had not escaped Charles Darwin with his need to conceive gemmules*:

> We cannot, in fact, consider orthogenesis as a mysterious internal tendency. But we are however obliged to note that since the origin of Life, we have witnessed an increasingly narrow adaptation of organisms to various biotopes […]. It is certain that we are witnessing, without being able to discern the exact cause, an increasing complication of organisms. On the other hand, it seems difficult to attribute this complication to the sole action of Darwinian factors such as natural selection or the struggle for life. The well-known facts of morphological convergence in groups of very diverse origins, but living in identical environments, seem to me to be difficult to explain, other than by some action of the environment, that is, by the intervention of a Lamarckian cause. (Arambourg, *Paléontologie, génétique et transformisme*, Commentaries, 1950, author's translation)

Teilhard intervened to show the reality of internal evolutionary drifts that were inexplicable by chance, the basic postulate that neo-Darwinism posed as necessary but, moreover, sufficient. He reported the case he observed in China, taking place over a period of 20 million years, that of the teeth of burrowing rodents, the Siphneus Spalacidae (mole-rats), originating from the same phylum but split into three branches distinguished by their skulls. The molars showed the same evolution "lengthening of crowns, loss of roots":

> which betrays a continuous process affecting the phylum itself, in its totality and in each of its components [...]. The Siphneus confront us with a certain case of directed morphological transformation, that is to say of orthogenesis. The phenomenon can no longer be denied. But the right explanation of the phenomenon remains to be found... (Teilhard, *Sur un cas remarquable d'un cas d'orthogenèse de groupe : l'évolution des Siphnéidés de Chine*, 1950, author's translation)

3.4. Hominization, "background orthogenesis"

These three orthogeneses, where nothing deviated in 20 million years, cannot be statistically explained by natural selection through random or accidental genetic mutations. The reiteration of three different evolutionary processes emerging from the same stem, over such long geological periods, defeated the probabilities. It was therefore necessary to consider internal processes more complex than the neo-Darwinian genetic synthesis. Teilhard de Chardin wrote that genetic mutations reinforced themselves, and he was very clear on the distinction between such physico-chemical processes and the metaphysical concept of finality. These were the premises of the sciences of self-organization or feedback effects.

> Followed for a sufficiently long time (palaeontological eras) the successive series of daughter-species thus engendered manifests the remarkable property of following the line of growth in value of a group of definite characteristics, the successive mutations not dispersing by chance, but reinforcing one another. There may be reason to consider continuous accentuation of certain characteristics (reinforcement of genes) quite outside any idea of "finality". (Teilhard de Chardin, *The Vision of the Past*, Cohen's translation, 1966)

The genetics of the 1940s was too incomplete in terms of molecular physics, and Teilhard was aware of this. It lacked the consideration of (1) electronic and atomic constraints in the combination of molecules assembled in macromolecules (chromosomes) and (2) electronic affinities between the varieties of molecules that make up the cytoplasm (the cell medium or metabolism). With regard to the electromagnetic experiments on the eggs of flies, they only demonstrated that the destruction of electronic bonds caused malformations, they did not explain the normal construction of a fly.

Moreover, ontogenetic processes are, authentically, the arrow of time or the duration, hence the evocation of Henri Bergson as written by Teilhard: the instant t1 contains its determination in t2, any organism is built with a succession of integrations of the preceding instant in a more complex state. It is this determination, acquired in the past, that is the object of the theoretical research on the transition between ontogenetic organizations when they form a phylogenetic succession of complexifications. Scientists describe these processes of the causal chain, with their trajectory: time is a parameter of the so-called deterministic equations.

The mechanism of evolution that takes place during gametogenesis can also be constructive durations (Dambricourt Malassé 1992). These hidden macroevolutionary processes, like the ontogenetic stages of any individual, are deduced from their fossilized ontogenetic effects, observed over very long geological periods. These so-called orthogenetic processes demonstrate, *a posteriori*, an orientation, or an iteration, and are characterized by emergences, or jumps, after long periods of stability. This is what made Teilhard say that macroevolution was a phenomenon of growth, irreversible and orthogenetic. Hominization was the continuation of this background orthogenesis, which had nothing to do with finalist, philosophical or metaphysical presuppositions.

To understand the order of an autonomous and animated molecular complexity in gametes, it was therefore necessary to start by understanding the formation of the very first unicellular cycles. However, "paleogametogenesis" was not on the research agenda and the amplification of an effect of gametogenesis on itself in the long term remained to be understood. It could begin to be explained by the so-called "autocatalytic" or "hypercyclic" processes, highlighted by the German biophysicist Manfred Eigen (1927–2019), Nobel Prize laureate in Chemistry in 1967. These are mechanisms of molecular synthesis that function in loops and which Teilhard was the first to call "self-coiling" at the cosmo-biogenetic scale.

> Such, for example, in the infinitesimal are the quanta, and relatively in the immense. Once we admit this, we have to ask what can be the specific effect proper to the vast complexes [...]. [T]his point of view [...] makes of biology simply the physics of the very highly complex [... Left to itself], without any doubt, one portion of the cosmic stuff not only does not disintegrate but even begins – by producing a sort of bloom upon itself – to vitalise. So true is this, that besides entropy (by which energy is dissipated), besides expansion (by which the layers of the universe unfold and granulate), besides electrical and gravitational forces of attraction (by which sidereal dust conglomerates), we are now

> forced […] to envisage and admit a constant perennial current of "interiorising complexification" that animates the whole mass of things. […] [T]he essence of true corpuscular complexity is that it appears in self-enclosed unitary groupings (unlike what happens, for example, in the crystal). Now, there are two different ways in which we may conceive such closed systems – according to whether they are found definitely arrested at their own level (as with a molecule of water or benzine) or, on the contrary, show themselves capable of modifying their composition, i.e. their complexity, without "unraveling" themselves (which is precisely the case with the cell). […] So far as actually existing is concerned, a cosmic movement of folding in upon itself seems quite undeniable. Where, however, are we to place its driving force?
>
> If our world is indeed a thing that is characterised by arrangement, then we can better appreciate that life can no longer be regarded as a superficial accident in the universe: we must look on it as (under pressure everywhere in the universe), ready to seep through the narrowest fissure at any point whatsoever in the cosmos – and, once it has appeared, obliged to use every opportunity and every means to reach the furthest extremity of everything it can attain: the ultimate, externally, of complexity, internally of consciousness. […] [M]an, the nucleus of a movement of in-folding and convergence in which, localised on our little planet (lost though it be in time and space) is manifest what is probably the most characteristic and most illuminating current that affects the immensities that envelop us: Man, on whom and in whom the universe enfolds itself. (Teilhard de Chardin, Man's place in Nature, Hague's translation, 1966)

Since its onset, the Princeton modern synthesis had lacked a reflection on the origin of the first unicellular organisms. Indeed, if there was one question that we needed to start with, in order to understand the principles of reproducibility of a process on long durations, it was the emergence of the unicellular cycle. Life is possible thanks to the chemistry of carbon and water molecules, but it is much more complicated conceptually than the appearance of unicellular organisms. Life is, at first sight, the appearance of processes organized in autonomous systems that draw upon their resources to reproduce within the environment in which they were "born". But *the conceptual difficulty is understanding how the process keeps its own form in its memory in order to reproduce.* Before taking into consideration the adaptation of a unicellular organism to its environment, it is necessary to understand this process, because natural selection is not the self-memorization of organized multimolecular dynamics, a*nd these processes capable of reproducing their own causality are the principles of orthogenesis.* This problem was not made explicit at the Paris Congress, but it was starting to be discussed.

Such dynamics had little to do with the Princeton doctrine, the genetic copying errors which would be the driving force of any, more complex, reconstruction of an organism under the baton of the "natural selection" conductor. The insinuations of finalism, vitalism and spiritualism were mainly aimed at making the young field of molecular genetics be admitted as the new inescapable science for the theory of evolution. But in substance, nothing had changed, it was still the same scale of Aristotle's beings ordered according to their complexity with the nervous system of *Homo sapiens* at the pinnacle, capable of consciously reflecting this natural history and situating itself in this history. This was what some scholars were trying to do at the Paris symposium at the *Institut de Paléontologie Humaine*. Evoking a hidden or unconfessed metaphysical presupposition was a misplaced argument in the circumstances, a tree hiding in the forest, faced with the objective reality of a macro-evolutionary problem, whatever the philosophical or religious convictions, as Haldane reminded us.

3.5. First synthesis: Man and the third axis of cosmic evolution of increasing complexity–consciousness

Pierre Teilhard de Chardin did not discover the process of this cosmo-biogenetic evolution, nor did he formulate the principles to understand this "self-coiling" mechanism, which was mainly a symbolic image. This process had reached the high level of molecular combinations that could be extended as "self-coiled", forming the first unicellular cycles on the Earth. This meant that the process had self-integrating properties, but this was precisely the un-thought-of problem, because this kind of process is not obvious. The question remained to understand how operating instructions could be their own operating instructions.

As a geophysicist he was familiar with scientific disciplines such as thermodynamics and then quantum mechanics. But it would be incorrect to suggest that he had discovered processes that would explain the emergence of the *Australopithecus* or the genus *Homo* from the gametes of the Miocene great apes. The only phrase in which he referred to evolutionary biology is in "the caprice of genes" that molecularists were beginning to identify.

Nevertheless, evolutionary paleoanthropology could not ignore his scientific and naturalistic synthesis, which sought to conciliate the increased complexity of the nervous system along with the emergence of consciousness that was increasingly reflected on itself. The natural place of Man remained that which Aristotle and specialists in comparative anatomy assigned to him and to themselves, the same as the first simians around 40 million years ago. This was not the top of a ladder, or of a creation, or of a divine plan, this place was measured for Teilhard, as it was for prehistoric paleontologists and paleoanthropologists, by the close relationship between the complex organization of the nervous system and the capacities for conscious reflection that emerged from it. At the threshold of *Homo sapiens* (Linnaeus 1758), the magnitude of the consequences of technical intelligence was that of a planetary phenomenology without equivalent in the animal kingdom. No lineage of the African great apes had left their continent, nor those of the *Australopithecus*, which became extinct, while the last stage, the *Homo* lineage, conquered the continents, the oceans, the stratosphere and then the solar system, in record time for a single species. And it was not bipedalism that enabled human societies to accelerate this intercontinental expansion, but the nervous system, with its cognitive capacities for anticipation and collective adaptation through cultural innovations.

Teilhard therefore did not approach the appearance of Man as a locomotor adaptation to an extension of the savannah over wooded environments, considering permanent bipedalism as a substitute for arboricolism. He did not use the expression "hominization" to describe an evolution of locomotion. He noted an acceleration of encephalization with the transition between craniofacial structural organizations. Integrating the work of Weidenreich, he noted the cranial capacity, but above all, the curvature of the skull and the degree of lowering of the cerebellar fossa: from the great apes since the *Dryopithecus*, the common craniofacial structural organization of *Homo erectus* in East Asia, Neanderthal Man in Europe and Rhodesian Man in Africa, and the last, *Homo sapiens*:

> The fundamental difficulty we come up against in the study of Evolution, once we reduce it to a process of cephalisation, is that we have not yet succeeded in defining the essential factor and hence the true parameter of cerebralisation – quite apart from the fact that should we ever succeed in defining it scientifically, *it will certainly prove to be a matter of neurons and not of osteology* […]. *It is true that by judiciously using and combining certain external indications empirically associated with internal advances in nervous organization, the increase in absolute volume and, still more, convolution of the cranium around its bi-auricular axis, we can follow in its main lines the development of the phenomenon* […]. (Teilhard de Chardin, *Man's place in Nature*, Hague's translation, 1966)

The volume of the brain is important, but less so than the arrangement of the neurons in the neocortex, which is unquantifiable, but nevertheless explicit with the increase curved shape of the skull that approaches Weidenreich's axis of rotation.

> This is sufficient to justify the conclusion that between the moment when we see the hominians attaining the Pithecanthropus stage and when they appear to us to reach their zenith at the Sapiens stage, they change, cerebrally, more rapidly and more profoundly than any other

> known living form during the same interval; even more, we may add, apparently, than the anthropoids themselves during the whole duration of the Miocene. *So important a biological fact cannot, obviously, be overlooked.* (ibid.)

Teilhard saw the Australopithecines as an extinct branch, with brains insufficiently complex in their organization to access the technical thinking necessary to produce a lithic industry. He did not speak of bipedalism for Australopithecines, but of a plantigrade locomotion and standing position. The origin of verticality was not discussed, even though it raised a conflict at the level of the articulation between the cerebellum and the spinal column: were they neural or locomotive in origin, were they internal phylogenetic in origin or an occasional environmental contingency? Teilhard avoided talking about this, as if it were a subsidiary question that would be unraveled later, its resolution not being able to call into question the background orthogenesis, the very nature of hominization, visible with the encephalization or the complexification of the neo-cortex with the formation of circumvolutions. This encephalization was thus the manifestation of the complexification of the nervous system since the first vertebrates, and was not astonishing since neo-cortical orthogenesis could be observed in other mammal lineages. However, what intrigued him was this originality that appeared as a spark in the human brain and that was missing in great apes, and, he believed, in *Australopithecus*: the capacity for abstraction in the design of tools, the shaping of pebbles and the production of shards. The few paleoneurologists of his time assumed that the volumetric difference of the brain, and especially the formation of neocortical circumvolutions, were related to these abilities.

The French paleontologist was a gradualist, but he distinguished two main types of processes, as we have seen. The first type of process did not ignore fossils because of the small number of populations of the stem species, those that appeared with a new organization, that thus were not yet specialized and were said to be generalized. The second process did ignore fossilized traces of the later radiation of this organization common to the varieties. They were classified in species, genera or families according to the age of the organization (the older it was, the more restrictive the nomenclators to increase the number of categories). This organization is still found today in as many ecological niches as allowed by adaptation mechanisms or specialization. The chimpanzee lineage is a divergent branch of the great ape stem population, whereas *Homo* is not a locomotor specialization of the oldest generalized great apes, but, as with those from which it derived, it forms a new ring of the trunk with encephalization and a lack of specialization. This new complexified grade again gave specialized and divergent branches across the surface of the globe, such as *Telanthropus* in South Africa, *Pithecanthropus-Sinanthropus* in

East Asia and Neanderthal in Europe. The last ring is *Homo sapiens*, the cradle of which he predicted to be in Africa, since, according to him, this continent was the soil that held the trunk.

For Teilhard, the internal organization of neurons in humans was the highest manifestation of orthogenesis since the first unicellular cycles. This cerebral organization being different, and more complex in *Homo*, in accordance with its less specialized cranio-facial structure when compared to the Australopithecines, the two lineages could not derive from the same stem population, and it seemed unlikely to him that *Paranthropus* could have made lithic tools.

In 1959, Louis and Mary Leakey continued their prospecting in Olduvai, Tanzania, in an area dating back 1.75 million years that was rich in lithic industries. They discovered a Hominidae skull contemporary with these archeological layers, giving the name to a new genus, *Zinjanthropus*, meaning "Man of East Africa". The morphology was comparable to the *Paranthropus* discovered in South Africa in 1948 in Swartkrans. Some scholars rejected their hypothesis that lithic tools had been produced by this Hominidae, because the brain was too small, with insufficient neocorticalization, but others maintained that it was plausible (Susman 1991).

Teilhard's innovative thinking was a phenomenology in the scientific sense, he observed the fossilized manifestations of disappeared processes and proceeded to a synthesis of knowledge that he put in order according to the scale of cosmogenesis up to biogenesis and neurogenesis. To the axis of space which tends toward the infinitely small and the infinitely large without zero, or center – as called to mind by Blaise Pascal, lost in a static cosmos – he added a second axis which was not only the time of the spatial expansion of the cosmos, but also the growth of complexity, starting from zero and that increased in size and complex arrangements: quarks, atoms, molecules, cells, organs and organisms (Figure 3.2). With these space–time coordinates, Teilhard drew his famous curve that followed the cosmogenesis in terms of energy and particles, with (1) a dotted line toward the infinitely small as it approaches zero – or the origins of the dynamic universe – and (2) an asymptotic incurvation which tends toward a spatial limit defined by the Earth's radius. The spherical limit of the globe forced the process to increase toward more and more complex organic structures, asymptotically. The nervous system of *Homo sapiens*, in which billions of combinations of nerve impulses are produced, matches with the last degree of this cosmic growth that found a matrix on the Earth. From then on, the neural networks of Sapiens would reflect information received by its sensors and think about them (how they did this remains a mystery). This is the phenomenon of reflexive consciousness, and this reflection makes up Aristotle's *anthropos*, the measurement of the World.

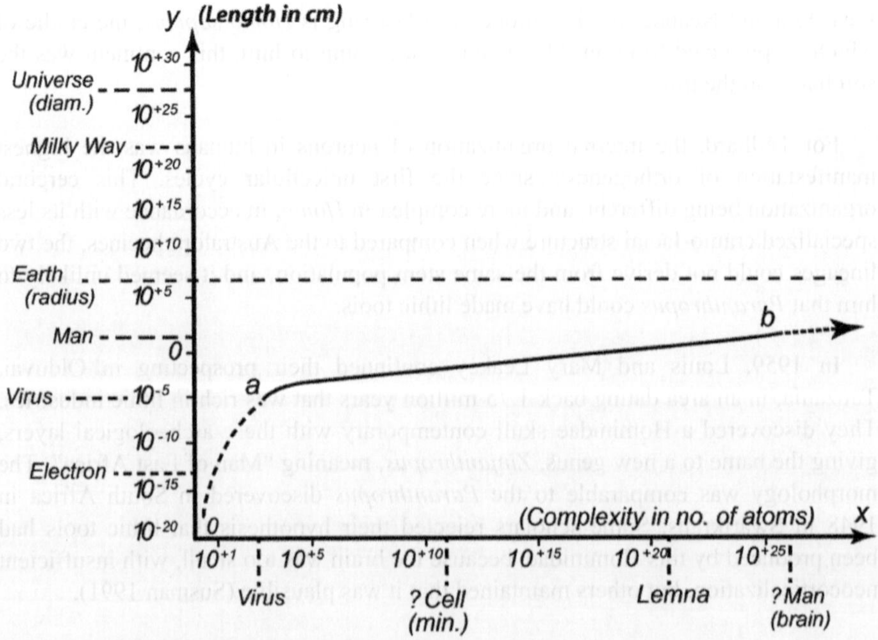

Figure 3.2. *Natural curve of complexities, a: point of vitalization, b: point of hominization (Teilhard de Chardin 1949)*

Teilhard's scientific synthesis is of an unequalled multidisciplinary scale and of an intellectual scope which exceeds the framework of descriptive paleontology. This space–time curve was traced according to new complexities due to atomic and molecular bonds. The growth of these complexities followed the curvature of the Earth, from the moist surface of the continents that moved through tectonics to oceans animated by warm and cold ocean currents. Teilhard included Man in these increasing complexities, by showing an organic, social and cultural phenomenon on a planetary scale, a super-organism that socialized and learned to understand reality in order to master and control its laws or anticipate their effects. These increasing complexities of the nervous system in different lineages of the animal kingdom thus formed not only a biosphere, but also a noosphere at the *Sapiens* stage. This neologism resulted from the exchanges between Teilhard and Edouard Le Roy (1870–1954), the successor to Henri Bergson (1859–1941) for the chair of Greek and Latin philosophy at the *Collège de France*. The geologist integrated "the atomic age" of the universe, which showed how much the unification of the cosmo-biogenesis had been accomplished in Man at all scales of organization, from quanta to neural and vascular networks from which reflexive consciousness emerged. This reflection was capable of writing equations like Schrödinger's, capable of controlling nuclear energy, or of dispersing it from the biosphere to the stratosphere. One must reach the threshold of *Sapiens*, or, more precisely, that of this noosphere, to see that hominization continues the cosmic processes that are highly integrated in the biogenesis. We then understand why Teilhard put anthropogenesis as part of planetology and, consequently, the relevance of the contemporary concept of the Anthropocene.

Life is a cosmic phenomenon, and what Teilhard meant by that was that at the threshold of the complexity of *Sapiens*' nervous system, its interactions with the biosphere and itself (the multiplicity of ethnicities and cultures) have reached a planetary magnitude. *Thus, the noosphere is the manifestation of a living cosmic phenomenon pushed to a very high degree of growth*. The great apes are not outside the process, but their interactions are not on a planetary and solar system scale, nor are they aware of the existence of the Earth as a planet; the orangutans do not know that their distant cousins live in Africa. They do not have a measurement of geological time. The emergence of reflexive consciousness is thus a new threshold of biogenesis, a cosmic phenomenon pushed to the extremes of its complex self-organization during gametogenesis, and from which neurogenesis emerged. Obviously, it was not the bipedalism of *Australopithecus* or of the first Man that constituted a cosmic phenomenon related to the complexification of quarks into atoms, molecules, then macromolecules with the footsteps of life on the Earth.

And obviously, if human anatomy should be interpreted as a locomotor specialization of the great ape, anthropogenesis cannot be understood as a cosmic phenomenon. It is then easy to bring up Anthropocentrism against Teilhard, a kind of retrograde blindness, because he was a Jesuit (Olivier 1965). But it was quite the contrary, his synthesis opened the eyes of intelligence by unifying the theory of evolution from pre-life to life, and made it possible to better follow what links reflective consciousness to cosmogenesis. His scientific knowledge and his spirit of synthesis allowed him to explore his own theological legacy, it was a personal essay. He did not approach straightening, *"the missing link" between the brain and the body*, convinced that there would be no contradiction with the background orthogenesis. But, above all, he brought a power of reflection that did not require theological doctrine to awaken human consciousness to its own phenomenological singularity, beyond all the characteristics that make it an animal like others: adaptations, varieties, malformations and spontaneous abortions. The reflection which becomes a consciousness of existence and of the Self, and thus conceives the Being, innovates with ethical considerations, existential preoccupations and paradoxical tenses between nihilism and the conviction to be someone in the view of the Other.

Teilhard was able to synthesize the scientific knowledge of his time and his curve was recognized by the scholars of his generation, such as the co-authors of the synthetic theory of evolution, George Gaylord Simpson and Theodosius Dobzhansky (1900–1975), president of the Teilhard de Chardin Association of New York in 1969, as well as Louis de Broglie (1875–1960), who reconciled the undulatory and corpuscular nature of photons and electrons, or the nuclear physicist Louis Leprince-Ringuet (1901–2000). His thinking can be seen in the lineage of Georges Buffon, Jean-Baptiste Lamarck and Georges Cuvier, even though he was not a transformist. The *Muséum national d'Histoire naturelle* at Paris supported the creation of a Teilhard de Chardin Foundation located in the *Jardin des Plantes*, within the walls of the Central Library. This foundation is a center for the documentation of his work and that of authors who have developed the approach of self-organized and autocatalytic processes. It has been recognized as a public utility by a decree of the Council of State since 1964. The first president was Jean Piveteau, who was succeeded by the current president, the prehistorian Henry de Lumley, former director of the *Muséum* and the last professor since Louis-Jean-Marie Daubenton to be elected to as director of the *Muséum* (1996–1999). Henry de Lumley is also the most recent holder of the *Muséum*'s chair of Prehistory, created in 1962, and has been President of the *Institut de Paléontologie Humaine* since 1981. Another member of the Teilhard de Chardin Foundation is Yves Coppens, holder of the Anthropology Chair at the *Musée de l'Homme* (1980–1983), then of the chair of Paleoanthropology and Prehistory at the *Collège de France* (1983–2005). During his inaugural lecture on December 2, 1983, Yves Coppens paid tribute to his predecessors, Henri Breuil, André Leroi-Gourhan (1911–1986) and Pierre Teilhard de Chardin, who should have held a chair in Evolution.

Certainly, Teilhard's thinking was too centered on the adult brain, the body was absent and bipedality and verticality were not considered, even though fossils from South and East Africa did open up the question. They propelled the evolution of the body and its locomotion into the problem of adaptation to ecosystems and terrestrial gravity, with, furthermore, different stages of the ontogenesis. As a consequence, the conditions for the development of intelligence from child to adult, with reference to Jean Piaget (1896–1980), were missing, as was the psychoanalysis of Carl Jung (1875–1961) and the notions of collective unconscious, of archetypes. But indeed it is Pierre Teilhard de Chardin who traced the main lines of a unification sought after for more than two millennia, between the cosmogenesis, tectonics, the complexification of Life and Man, as an organism capable of reflecting consciously on all this natural history. It is therefore inconceivable in France to be unfamiliar with Teilhard's thinking in evolutionary paleoanthropology, even though he was not an anthropologist, once the fundamental research addresses the processes of hominization. The creation of the Teilhard de Chardin Foundation marked the recognition of a threshold in the theory of evolution that embraced the *Muséum national d'Histoire naturelle*, the *Collège de France*, Sorbonne University and the

Institut de Paléontologie Humaine with a conciliation between cosmogenesis, biogenesis, anthropogenesis and the emergence of the self-reflexive consciousness, in search of its terrestrial and cosmic identity.

From this point in time, this reflection went through the gametogenesis of *Homo sapiens*, a verticalized body before birth, not the axis of the world as in Aristotle's time, but the current stage of an ongoing process and one which was distinguished from other lineages by the emergence of reflexive consciousness, by the capacity to think about the evolution of the universe from its beginning to the fossil and current terrestrial species, by the capacity to include oneself in it, as the only animal responsible for its own actions, anticipating the consequences in the long term, through action or inaction. *Homo sapiens* must also look after itself, and discern in posture problems linked to verticality and those of dental occlusion (with or without agenesis) the possible signs of this ongoing evolution which act on the complexification of the nervous system from its embryonic formation. This study can only be based on a comparison with the fossil species in order to discern the modalities of this "hominizing phenomenon" and to be concerned about what we are becoming. We will see that, if bipedalism is terrestrial, verticalization with its psychomotor implications is definitely part of the cosmic curve of self-complexification.

4

The Body, Arboricolism and Adaptation: The Years 1950–1980

4.1. Under the brain, a body

The evolution of the human lineage was already well publicized in the 1960s and 1970s. The *Time Life Nature Library* published Ruth Moore's "Evolution" volume in 1962 and 1964, with watercolors by Maurice Wilson (1914–1987). One of these watercolors is a semi-erect *Proconsul*, clinging to a trunk in a richly wooded savannah with a Rift Valley volcano in the background. Another watercolor shows a South African *Australopithecus*, with an erect human body with a "monkey-man" head and a stooped neck, looking down on a savannah with sparse trees, well sheltered in the crevices of a karst massif. Teilhard de Chardin's recognition of Africa as the cradle of hominization is presented as the surest of validations. The drawings illustrate the external appearance of the adult skull, always giving the illusion of a gradual transformation from one adult to another. Nothing is said about the mechanisms that could have transformed the morphology of the pelvis or that of the mandibles.

Hominization was taught in colleges with a theory that was still famous at the time, that of the Dutch biologist Louis Bolk's (1866–1930) theory of fetalization. He formulated it in 1924 when embryology and differences in growth rates (the heterochronies of development) were fields of research in evolutionary biology. This theory was in force in the teaching of Jean Piveteau (1963) and based on the volume ratios between the endocranium and the face. Indeed, the fetal stages of the chimpanzee and gorilla show a high volume ratio (a large brain for a small face) which is reversed after birth, when brain growth ceases in favor of the face, which becomes prognathic. Then, the sphenoidal angle opens, as Herman Welcker had already observed in 1862 (see Volume 1). In comparison, Man's volume ratio became accentuated during development and never reversed, so the sphenoidal angle hardly opens at all. Bolk therefore hypothesized that hominization is the complexification of the nervous system, essentially that of the two cerebral hemispheres from the time of fetal growth, with the effect of maintaining the straightened position of the clivus with the occipital hole lowered forward. In fact, he thought he saw this position in a canid embryo and deduced that it is the first in vertebrate embryos (Figure 4.1). According to this hypothesis, the human position is the most primitive and has been preserved because of the pressure exerted by brain growth on the cerebellum, whereas this pressure has ceased in monkeys, freeing the cranial base from these constraints. According to him, hominization *is* fetalization and above all, *the straightening is not the evolutionary process since the prosimian strain*. Yet, microcomputed tomography of mice never showed a straightened clivus (Xiaoyou *et al.* 2017). The description of the canid embryo is only a diagram.

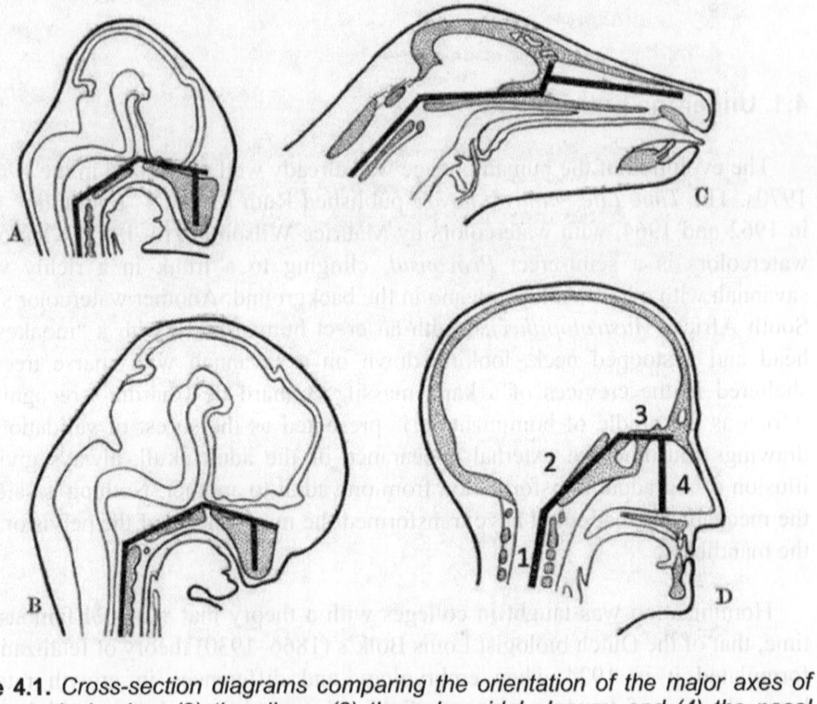

Figure 4.1. *Cross-section diagrams comparing the orientation of the major axes of (1) the cervical spine; (2) the clivus; (3) the sphenoidal planum; and (4) the nasal septum in the embryonic and adult stages of a canid (A and C) and modern Man (B and D) according to Louis Bolk (Piveteau 1963)*

Jean Piveteau did not make a final decision on this topic and so taught both hypotheses, the neural cause and the locomotor cause. Fetalization was on the right track for the brain and the face. There was no doubt about the intrauterine origin of hominization, whose most spectacular indicator is the increase in convolutions from the fetal period, while the face is clearly less developed than that of an already prognathic Pongidae. However, fetalization did not explain the neo-natal differences visible on the postcranial skeleton. We will also see that the schematic diagram of the dog's embryo does not reflect reality, because the skull base is never flexed; this fact is responsible for a lot of confusion in the first quarter of the 21st century.

In 1968, new geological terrains were prospected in the Afar region of Ethiopia, where the Rift Valley joins the Red Sea at 274 m below sea level, in this great collapsed trough (graben) that starts the formation of an ocean. The cartography was drawn up by Maurice Taieb, for his doctoral thesis in geology at the Pierre and Marie Curie University – Paris VI. Taieb located numerous Pliocene and Pleistocene (3.2–2.3 Ma) fossiliferous sites in the Hadar region and set up a prospecting and excavation program, the International Afar Research Expedition, in 1972. The French team was co-directed with Yves Coppens (Coppens 1975, 1980), then senior lecturer at the Paleontology chair of the *Muséum*, which was directed by Jean-Pierre Lehman (1914–1981) from 1956.

In 1974, several Austrlopithecus fossils were collected, including an almost complete skeleton, by the international team in the Ethiopian concession attributed to the American team led by Donald Johanson. The mandible is the part of the skeleton that is most frequently preserved and is therefore over-represented. The new species, named *Australopithecus afarensis*, began to attract attention, not only because it was found thousands of kilometers from the South African sites, but also because it was older than the *Australopithecus africanus* species and the *Paranthropus* genus. But the great novelty was the traces of arboricolism with the high position of the scapula, the elbow joint and the length and shape of the phalanges. The bones of the foot also had a joint at the level of the hallux (thumb), which allows for prehension, like the feet of the great apes. The forelimbs, on the other hand, were no longer than the lower limbs, so these *Australopithecus* species, although bipedal on the ground, must have lived in trees.

The oldest Hominids were arboreal, so how can we justify their axial skeleton, transformed and straightened into a bipedal locomotion exclusive to the ground?

The *Australopithecus* radiation extended beyond the equator in the northern hemisphere, so the search now extends to the entire rift, from Ethiopia to Malawi, Kenya and Tanzania in search of the emergent cradle (Figure 3.1).

The use of exclusive bipedalism on the ground, deduced from skeletal anatomy, was supported by the footprints of Laetoli (Tanzania) dated 3.66 million years ago, formed on soil covered with soggy ashes. The conformation of the footprints was not that of a human, as demonstrated by Deloison (2004), who highlighted that there was an absence of a plantar arch and the presence of a powerful first toe retractor muscle (musculus abductor hallucis brevis). The thumb was prehensible and the imprints compatible with the morphology of the foot bones of the Afar fossils. Chimpanzees leave similar imprints, and since they occasionally adopt bipedalism, the footprints of Laetoli raised a doubt, but they remain a strong argument, because not only are *Australopithecus* already known in Tanzania with a mandible in Laetoli and a maxilla in Garusi, but such prints would be left by the feet of arboreal *Australopithecus*. It was not so much the bipedality as the straightening of the central nervous system that was important, from the sphenoid to the sacrum. As is often the case, the bases of the skulls were missing. The *Australopithecus* was thus quickly reduced to one and the same problem; locomotion imposed by a hypothetical radical change in their environment. Then, there appeared expressions such as "mosaic evolution", which make it even more confusing to understand the coexistence of arboricolism and movement in bipedal mode, placing walking and arboricolism at the same evolutionary level. Arboricolism is a function adapted to the way of life, whereas walking is the locomotion imposed by the straightened structure of the axial endoskeleton before birth, straightening being the consequence of the phylogenetic process of complexification, which we will once again address.

Coxal bones of young *Australopithecus* were discovered at Sterkfontein (Sts 14) in 1949 and at Makapansgat (MLD 7) in 1972. Their morphology already shows the short and broad structure from which, and within the limits of which, growth and learning of the bipedal posture would take place until the adult stage. Everything already showed an evolution of the morphogenesis of a great ape, the theory of fetalization still supported the comparison for the brain and the face, but not for the post-cranial skeleton.

The *Muséum national d'Histoire naturelle* (Paris) thus participated in the first paleoanthropological studies on the *Australopithecus* of the Afar, under the direction of Yves Coppens, focusing on research on the locomotor skeleton, with Yvette Deloison on the foot, Christine Tardieu on the knee, Christine Berge on the growth of the pelvis and Brigitte Senut on the shoulder and forelimb. The mandible was not a priority subject of study; no one imagined that it could shed light on the completely original locomotion of *Australopithecus*, plantigrade and arboreal. The cerebellum was reduced to its function of controlling postural balance, and the mandible to its manducatory function, essentially masticatory. The brain thought a little, the cerebellum controlled the posture automatically, the jaw chewed, but the function of the base of the skull remained unknown, conceptually stuck at an extremely confusing crossroads between the two processes defined previously by Teilhard de Chardin, the evolutionary change of the morphotype or just a specialization among others. Either the origin is neural and allows us to go back up the thread to the genetics of the fertilized egg, or it is a locomotor constraint, but since the discovery of molecular genetics, it is no longer a question of the hereditary transmission of the characters hypothetically acquired after birth, with the locomotor function. The "mosaic evolution" is not an observation that gives an account of the reality of evolutionary processes, it is only a phrase used due to the lack of distinction between the phylogenetic increasing complexity channelized on the nervous system (macroevolution) and the radiation or environmental adaptation, such as arboricolism (microevolution).

4.2. Hominization of the skull and posture, French schools

4.2.1. *The* École de Paris, *The Sorbonne Museum of Natural History*

Dr. Robert Gudin's pantograph (1951)

The first unification of the endocranial base with the face was in the work by Dr. Robert Gudin (1911–2010), who was a speech therapist connected to the Faculty of Medicine of Bordeaux, who specialized in disorders of the oro-facial sphere. This unification was a geometric drawing constructed in 1951, in the framework of a doctoral thesis entrusted by Jean Piveteau, and defended at the Sorbonne under the presidency of Pierre-Paul Grassé. Pierre Teilhard would have had time to read it and attend his thesis defense, if he had been in Paris, and not in New York. Dr. Gudin's thesis consisted of demonstrating the benefit of lateral teleradiography to better study these disorders of the ENT (Ear, Nose and Throat) sphere, which involve the back of the throat and thus the pharyngeal (external) face of the clivus. The speech therapist had perceived the geometrical links between the orientation of the clivus and the degree of projection of the face containing the nasal cavity. He made X-rays of skulls of different ethnic groups from the collections of the *Musée de l'Homme* and the Comparative Anatomy collection from the *Muséum* with (1) prosimians *Archeolemur, Propithecus* and the Tarsier, (2) a small American monkey with Ateles (spider-monkey), and (3) two great apes with the chimpanzee and the gorilla.

His profile analysis traced a triangle (Figure 4.2) whose summit represents the degree of straightening of the clivus and the sliding of the face under the anterior cranial fossa, that is, under the frontal lobes. The triangle closes following the classification of primates from the most primitive to *Homo sapiens*. Gudin came to the conclusion that the skull is a dynamic system in which the face and its functions are dependent on the statics or the balance of the axial posture. The sphenoid was the key to reading the dynamic center that unifies the face and the orientation of the spine:

> The movements of the face in its phylogenetic evolution and in its development in relation to the skull: to understand this regression of the face under the skull, it is necessary to study other angles (other than the sphenoidal angle and the angle of facial prognathism). If we analyze the behavior of the facial mass in relation to the

> sphenooccipital plane, it emerges that the angle (of the pantograph) gives a series of variations that depend on the static of the individual and the functional face-cranium balance in relation to its degree of development. (Gudin, *Étude du crâne de profil. Morphologie et radiographie*, 1952, author's translation)

A few years later, Gudin integrated the mandible, retaining the inclination of the central incisor with its root and that of the ramus, whose condyle is hinged with the part of the skull base which straightens. In the end, his analysis drew a pantograph, and made the dynamic relationship expected from the model explicit (Figure 4.2): a straightened clivus corresponds to a symphysis and a maxilla with little prognathism (Gudin and Ducasse 1973). His thesis was awarded a prize by the French National Academy of Medicine (*Académie nationale de médecine*) in 1952. Gudin's analysis has become a reference in France for the early treatment of malocclusions, mispositionings of the maxillo-mandibular osseous matrices and of the skull base bones. Since then, it has been corroborated by cephalometric analyses applied in infantile dentofacial orthopedics (Deshayes 1986).

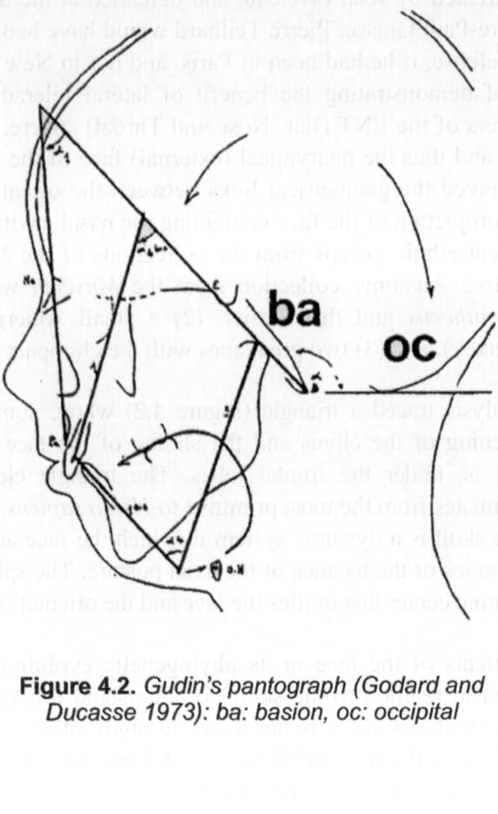

Figure 4.2. *Gudin's pantograph (Godard and Ducasse 1973): ba: basion, oc: occipital*

As Franz Weidenreich deduced, modeling orthopedic treatments by using lateral teleradiography and the pantograph confirmed the presence of an axis of rotation in the center of the skull base. There are therefore forces that are in action somewhere in the sphenoid whose vectors follow rotations, but the location of the first transverse axis, and especially the origin of the forces, is still unknown. The answers involve the sutures of the base, composed of three cartilaginous territories (chondrocranium), the ethmoid, sphenoid and basi-occipital, separated by synchondroses, the spheno-ethmoidal, intra-sphenoidal and spheno-occipital (Figure 4.3).

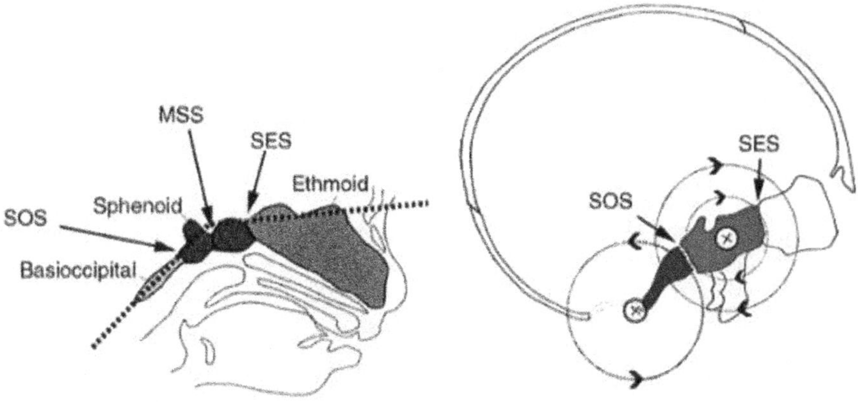

Figure 4.3. *The territories of the Homo sapiens' chondrocranium, mid-sagittal section. Left: a fetus from Lieberman and McCarty (1999), theoretical approach, without rotation vectors. Right: an adolescent according to Dr. M.J. Deshayes (1986), clinical and dynamic approach with rotation vectors. MSS: middle sphenoidal synchondrosis; SES: spheno-ethmoidal synchondrosis; SOS: spheno-occipital synchondrosis. Rotation vectors are made possible by the sphenoidal synchondroses, as long as they are not synostosed*

The principle of the dynamic pantograph imposes several axes of rotation. These vectors have been recognized by the Sorbonne since 1951 thanks to Robert Gudin, who brought understanding of the relations between the base and the face to the forefront of French schools. *This avant-garde has become that of human paleontology with the integration of vectorial forces and the evidence of internal dynamic logics revealed by the modeling of pantographs.*

The living mechanics of André Leroi-Gourhan

The ethnologist and prehistorian André Leroi-Gourhan held the chair of Prehistory at the *Collège de France* between 1969–1982. Jean Piveteau had entrusted him with a thesis in the continuity of Robert Gudin's, entitled "*Tracé d'équilibre mécanique du crâne des Vertébrés terrestres*"[1], which complemented his own conclusions; the thesis was defended in 1954 and published 30 years later (Leroi-Gourhan 1983) (Figure 4.4).

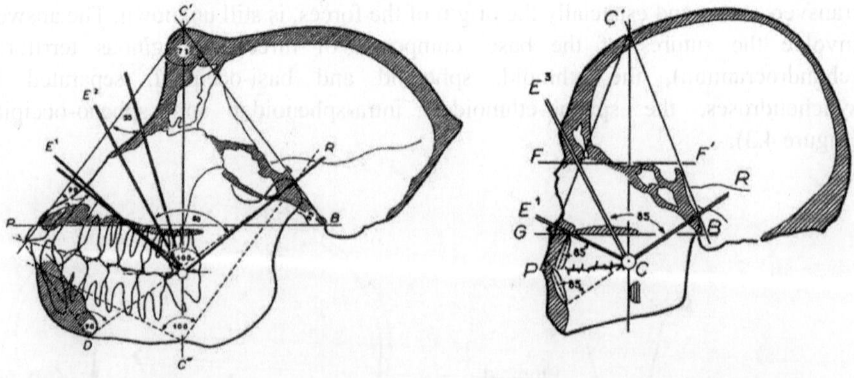

Figure 4.4. *Geometric model of the basi-craniofacial frame and the angular correlations between the posterior base and the face, applied to the adult gorilla and Homo sapiens. Taken from the thesis of Leroi-Gourhan (1983)*

If there is an absolutely formal connection between the vertebral axis and the basilar floor of the skull, this connection is independent of variations in the volume of the brain. It thus appeared *that one of the theoretical lines most frequently followed by anthropologists, that of the action of cerebral dynamism on the evolution of the cranial base and attitude, was contradicted by the most diverse paleontological materials*. A second general phenomenon has been identified, the absolute constancy of a relationship between the basion (anterior point of the occipital hole) and the prosthion (bony point between the incisor alveoli):

> *any decrease in the length of the posterior skull leads to an equivalent decrease in the dental skull front.* The inclination of the basilar clivus is directly related to the acquisition of vertical station, but it also seems that the cranial base (station) and dentition respond proportionally in higher mammals. In primates, the length of the dental device corresponds exactly to the mid-distance between the prosthion and the basion, which confirms the interplay between station and posture. From this state, the problem appears as a construction problem. A final fact that has proven to be important is *the response of the inclination of the mandibular symphysis and the basilar clivus,*

1 Translation: Traces of mechanical equilibrium in the skulls of terrestrial vertebrates.

which creates a system of relationships between the dental device and the base. (Leroi-Gourhan, *Mécanique vivante*, 1983, author's translation)

4.2.1.1. *Antoine Delattre's* École de Lille *(1950–1960)*

Antoine Delattre (1891–1983) held the chair of Pathological Anatomy at the Faculty of Medicine of the *Université Catholique de Lille*. In 1924, he defended a natural sciences thesis at the Sorbonne, entitled "*Essai sur l'anatomie comparée et la mécanique fonctionnelle de l'axis des Mammifères*"[2], under the direction of Raoul Anthony (1874–1941). Raoul Antony was a passionate Lamarckian and had a profound influence on Antoine Delattre, who founded the *Laboratoire de Craniologie Comparée de Lille*[3] in 1950, before it became a research unit of the CNRS in 1969. He was president of the *Société d'Anthropologie de Paris*[4] in 1957. With his collaborator, Dr. Raphaël Fenart, Delattre devoted 10 years of research to the mathematical modeling of skull hominization thanks to the development of teleradiography. The horizontal was given by the lateral semicircular canal of the inner ear located in the petrous pyramids. The vertical reference crosses it at the center of the canal. These models have been applied to the adult skulls of current carnivorous and herbivorous species and to all categories of primates in the *Muséum's* Comparative Anatomy collections. The comparisons show no lowering of the cerebellar fossa in non-primate vertebrates and prosimians; the lowering increases by stages from monkeys to *Homo sapiens*. Well aware that the phylogenesis of all these current species relates to their ontogeny, Delattre and Fenart compared the growth trajectories from birth to adult in gorillas and *Homo sapiens*. They showed the paths of the vault, the clivus and the occipital hole, which are rotations around point zero. They were said to be positive that the trajectory followed from top to bottom and tended toward verticality, and negative in the opposite direction. The plicature of the base is always positioned at the level of the sphenoid – that is not new. But, both authors believed the movement was driven from the occipital and they called it the "occipital bascule". *The sphenoid was not considered the center of the first movement, probably because of the preponderance of the Lamarckian school, which was extremely prevalent at the Sorbonne and the Muséum, and focalized on the occipital hole and the vertebral column since Daubenton and Lamarck.*

Indeed, Delattre and Fenart estimated that, at birth, the occipital bascule of the gorilla and *Homo sapiens* are identical, but continues in Man due to bipedal locomotion. However, in humans at birth, the sphenoidal angle is actually more

2 Translation: Essay on comparative anatomy and functional mechanics of the mammalian axis.

3 The Lille Laboratory of Comparative Craniology.

4 The Anthropology Society of Paris.

closed and the occipital hole is more advanced than in the gorilla (see Figure 5.36), so locomotion cannot be the cause of a more straightened clivus. Moreover, the term "hominization" is an extrapolation since the authors had not studied the ontogeny of fossil species that preceded *Homo sapiens.*

4.2.1.2. *The* École de Nantes *and the cephalometric analysis of Jean Delaire*

Jean Delaire was born in 1923. He became a Doctor of Medicine at the University of Nantes in 1953 and specialized in stomatology in 1957. Then, he became head of the Department of Stomatology and Maxillofacial Surgery at the University of Nantes in 1959. As early as the 1970s, he became concerned with children suffering from mandibular prognathism, or promandibulia, and developed an orthopedic mask which established the international reputation of the Nantes school. Jean Delaire was an honorary member of the British Association of Oral and Maxillofacial Surgeons and the Canadian Society of Oral and Maxillofacial Surgeons.

Orthopedic exercises involve correcting malocclusions, that is the misalignments between the two dental arches when they are in contact. Delaire defined a cephalometric analysis that associated the base and the face in profile, traced on teleradiographies (Delaire 1978; Lalauze-Pol *et al.* 2009) that was completed by Deshayes (1986) (Figure 4.5). Orthodontists distinguish three classes with a standard, or class I, defined by the position of the lower first molar relative to its maxillary homologous. The anterior cusp of the mandibular molar must be behind that of the maxillary molar. When the occlusion is balanced, the bony structures of the mandible, maxilla and the skull base have a geometrical or angular relationship that corresponds to a physiological balance, such as good swallowing and nasal breathing. Malocclusions are said to be in class III when the lower molar is forward of its maxillary counterpart and the joints between the mandible and the skull base (the temporomandibular joint, TMJ) are shifted forward. They are said to be in class II when the molar is moved backward and the TMJs are too high. Traces made on lateral teleradiographies show that the imbalances are correlated with the orientation of the clivus divided into two territories by the spheno-occipital (or sphenobasilar) synchondrosis, considered as the center of the basicranial flexion.

To the bony tissues, Delaire added the dura mater, which lines the inner surface of the skull, separating the two cerebral hemispheres with the cerebral falx, then the occipital lobe and the cerebellum forming the tent docked on the crest of the petrous pyramids and the occipital, before extending into the vertebral column. This bracing allowed him to correlate angular variations in the two sagittal and transverse planes of the base most directly concerned with verticality, namely, the petrous pyramids (Delaire 2003). Their great axis converges forward toward the sphenoid and intersects in the plane of symmetry, forming an angle called the "bi-petrous angle".

An accentuation of the positive rotation would lead to the opening of this angle, the pyramids tend to be parallel with the frontal (or coronal) plane, this trajectory has the effect of moving the TMJs away from each other and thus of moving the alveolar-dental arch away:

> His contribution to orthodontics is enormous. The more the sphenobasilar synchondrosis is flexed, the more the bi-petrous angle is open, which implies an advancement of the mandible and favors skeletal class III, and conversely, the more the sphenobasilar synchondrosis is in extension, the less the bi-petrous angle is open, and this leads to a retreat of the mandible and favors skeletal class II. (Benoit-Levy, *Théories plus récentes* (Benoit-Levy n.d.), author's translation)

Terms such as "extension" refer to dynamics, to forces of action, which reflect the break with the dogma of the genetic pattern, or the metaphor of the genetic "program". The demonstration of active forces in synchondrosis, and SOS in particular, is based on the success of orthopedic treatments that consist of correcting growth vectors with masks connecting the vault to the dental arches. These masks transmit millimeter-scale tensions or pressures, either by adding stress in the direction of flexion or by opposing resistance to it. After several months, the occlusion may return to class I equilibrium. *The treatments are not chemical and the genes do not express resistance, rather, they are solicited.* Dynamics therefore refers to the force vectors that act in the synchondroses and give the main directions of growth to the various related cartilage units. Obviously, the forces acting before birth cannot be assigned to locomotion nor chewing.

The pyramidal frontalization (i.e. the opening of the bi-petrous angle), caused by the basicranial flexion, is a particularly important clinical finding for human paleontology in order to understand *Australopithecus* and the early fossil species that define the genus *Homo*, as paleoanthropologists still believe that cerebellum volume is the driving force. ENT research has brought enlightening knowledge on the complexity of these repositionings that do not follow linear vectors (growing sideways and pushing down and forward). Indeed, axes of rotation are themselves driven by other axes of rotation. The example is the temporal bone considered as a whole with the petrous pyramid and the squamous that forms part of the temple. Its trajectories are comparable to a wheel driven by its rotating axis. The long length of the pyramid is the axis of rotation, the squamous is the wheel. The displacements are then compared to a veiled wheel. If the rotation is too pronounced, the bi-petrous angle is open and the temples tend to move away from the skull. Conversely, if the flexion is insufficient, the temples are compressed (Hadjouis 1999, 2018, 2021; Cho 2005). What is difficult is identifying the motor of the pyramidal rotation by looking for it in the spheno-occipital synchondrosis and probing it, in order to understand at which scale the tensions are located, the nature of the forces that are exerted there and the role of the fascia (connective tissues).

4.2.1.3. *The cranio-palatal balance of Dr. Marie-Josèphe Deshayes (1986)*

Jean Delaire's analysis was taken up by the French stomatologist Marie-Josèphe Deshayes, Doctor of Medicine at the *École de Nantes*. Her clinical research refined cephalometric analysis and developed a physiological and architectural principle called "cranio-palatal balance". Our approaches collided in January 1988 at the *Musée de l'Homme* during the annual congress of the *Société d'Anthropologie de Paris*. It was the first ever presentation of the "craniofacial contraction" process deduced from our study on mandibular ontogeny of current and fossil primates (Dambricourt Malassé 1987). Jean Piveteau also presented at the Academy of Sciences in April 1988 before his presentation was published in the *Comptes Rendus de l'Académie*, entitled "Hominisation et foetalisation" (Dambricourt Malassé 1988). We collaborated for roughly 10 years in the continuity of the *École de Paris La Sorbonne-Musée national d'Histoire naturelle*, initiated in the French school thanks to Jean Piveteau with Robert Gudin and André Leroi-Gourhan:

> The author wonders why so many children need orthodontic treatment. She cites a 1998 report by the French Union for Oral Health in which it is said that out of 1470 children, aged 7 years, 70% need orthodontic treatment. M.J. Deshayes is trying to understand cranial osteopathic theories, which she calls cranial biodynamics. For M.J. Deshayes, the cephalic morphogenesis is done by a more or less opposed flexion program of the sphenobasilar synchondrosis and by local morphomechanical factors explained by osteopaths. It advocates early orthodontic treatment, i.e. before the age of 6. (Benoit-Levy n.d., author's translation)

Cranio-palatal balance is defined by a set of straight lines and angles that are a function of the sphenoidal angle, and the homologous points that enable them to be drawn concern the occipital squamous, sphenoid, ethmoid, mandible and maxilla (Figure 4.5). Marie-Josèphe Deshayes has shown how *this balance is already visible at the fetal stage, before dentition, so it is not an occlusion-constrained architectonics. The classes are therefore not dental, but skeletal*. The right balance, or skeletal class 1, is defined by a set of angles fluctuating around an average. Imbalance occurs when angular fluctuations exceed amplitude thresholds. The functions are thumb-sucking from the fourth intrauterine month, as well as swallowing. Dentition is therefore set up in this context of balance acquired before the post-natal functions of chewing, breathing and locomotion.

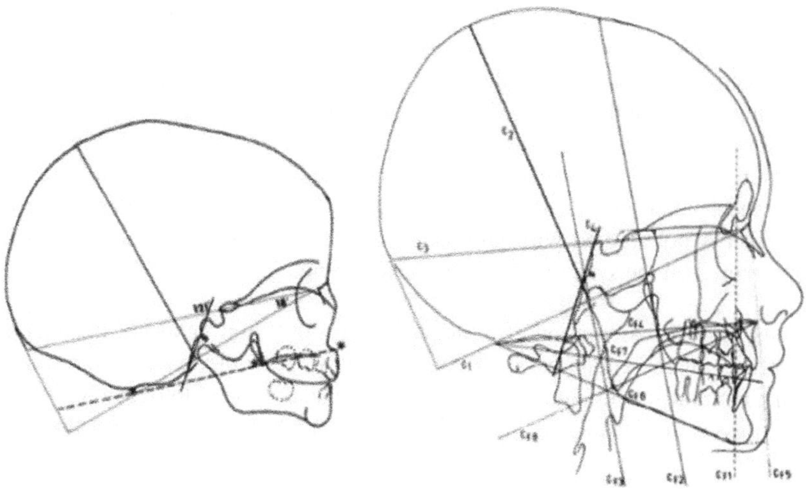

Figure 4.5. *Cranio-palatal balance of Homo sapiens in the neonatal stage and in mixed dentition (Deshayes 1991). The cranio-palatine line corresponds to the alignment of the nasal spine (asterisk) with the contact of the palate and the pterygoid apophysis of the sphenoid (asterisk). Balance is found when the alignment is tangent to the lowest part of the occipital squamous (asterisk). This line is visible in mixed dentition with the craniofacial line 4, or CF4. The cranio-palatine line is parallel to the spheno-ethmoidal plane. For a color version of this figure, see www.iste.co.uk/dambricourt/embryogeny.zip*

The analysis is applied to children before the placement of the first permanent molar when the synchondroses are not ossified, in other words, before the start of mixed dentition (both temporary and permanent). The morphodynamic uniqueness of the organism is allowed by the fascias, up to the intra-sphenoidal synchondrosis composed of collagen fibrils (Dambricourt Malassé 1989) (Figure 5.19). The task consists of evaluating the dynamic at the level of the synchondrosis responsible for the imbalances. Some children have an occipital squamous lower than the cranio-palatal balance, in this case the sphenoidal angle is closed and the mandible is propelled. Conversely, other children have an occipital squamous above the CF4 line, in this case the sphenoidal angle is open and the mandible is recessed. Orthopedic and orthodontic treatments aim to restore the cranio-palatal balance, so as to harmonize the occlusion between the two dental arches. Deshayes also notes the opening of the bi-lateral angle with a large flexion in class III (Deshayes 1986, Figure 48). The treatments are extra-oral forces that slow down or activate flexion at spheno-occipital synchondrosis. These corrections are accompanied by a modeling of the skeletal structures of the face, preferably in the vertical direction in insufficient flexion, or, conversely, in the antero-posterior direction in the case of great flexion.

If the vectors are not those of a basicranio-facial pantograph, then the return to dental articulation in class I is not observed, and so the orthodontic treatments would be ineffective. These forces therefore act on the synchondroses through the supporting tissues, either in the direction of a contraction if the whole is in class II (extension), or by slowing down its contraction if it is in class III. Finally, these principles demonstrate that neither the mass of the brain, nor that of the cerebellum, are the mechanical causes acting on the sutures of the base. The basicranio-facial contraction is not studied with the fascias of the vault, but with those of the skull base, centered on the synchondroses of the sphenoid.

Thus, clinical observation of orthopedic constraints highlights correlations in the three planes of the endocranial space of the base, with the axial posture, but also with the temporo-mandibular joints and thus, the occlusion (Motoyoshi *et al.* 2002). These stresses are located on the surface of the cerebellar fossa and at the level of the spheno-occipital synchondrosis. Their nature and origin remain a major concern in dento-maxillofacial orthopedics and are the subject of debate, since they are neither measurable nor directly observable on the patient, unlike the exoskeletal biomechanical forces exerted by muscle tone (e.g. Bergamini *et al.* 2008). These internal forces are completed by external muscular forces such as the tongue, which interacts through pressure on the incisor arch and can activate the return of the dental joint in retromandibular. Stomatologists and orthodontists refer to the "lingual guidance" as one of the mechanical factors contributing to the occluso-postural balance.

There is clearly a systemic and dynamic axial unit around the central nervous system and its fascias, which is first in the order of basicranio-facial imbalances, centered on the sphenoid and, therefore, likely since organogenesis.

5

The Embryonic and Phylogenetic Origins of Human Posture

5.1. A reversal of perspectives

5.1.1. *The origins of Man: the first doctoral school of the* Muséum national d'Histoire naturelle

The *Muséum national d'Histoire naturelle* acquired University status in 1981, 355 years after the creation of the *Jardin royal des plantes médicinales*. This page in the *Muséum*'s history was the work of the prehistorian Henry de Lumley, following his election as the chair of Prehistory. Trained notably by Henri Breuil and Jean Piveteau, the latter wrote about his thesis, defended in 1965:

> One of the capital contributions of Henry de Lumley's work is the reconstitution of the social structures of prehistoric Man. From the Hominids who lived in the Vallonnet cave until the advent of *Homo sapiens*, we see the aspects of social structures becoming clearer and more complicated, as well as the successive stages, over a long period of time, of the phenomenon of hominization. Only prehistory can enlighten us on the evolution of this social fact, which will allow us to understand the human genesis and the scope of the human phenomenon when it reaches and then goes beyond the stage of thoughtful thinking. (Piveteau 1969, author's translation)

Henry de Lumley was at the head of the largest French research unit specializing in prehistory, uniting the research laboratories of the *Muséum national d'Histoire naturelle*, the Faculty of Medicine of the University of Aix-Marseille, the *Institut de Paléontologie Humaine* (IPH), Tautavel, Le Lazaret and Terra Amata. This federation represented the largest collection of Hominidae fossil casts in the world, to which the original *Homo erectus* sensu lato, Neanderthal and Cromagnoid fossils were added. Thus, in 1981, the *Muséum*'s first doctoral school entirely dedicated to the origins and evolution of hominids in their environmental, cultural and social context was created. It was called "Quaternary Geology, Human Paleontology, Prehistory".

5.1.1.1. *As a prologue*

This school welcomed the research theme that I wanted to develop on Louis Bolk's fetalization theory, starting in 1982 for a period of five years, after first completing a bachelor's and master's degree in Earth Sciences at Paris XI-Orsay University (Paris-Saclay). This was the first thesis in human paleontology belonging to the *Muséum national d'Histoire naturelle*. The objective was to clarify the paradox between the theory of fetalization, *a priori* relatively coherent for the head, and the post-cranial skeleton of *Australopithecus*, considered from the locomotor perspective.

Beneath the brain was a body, too often neglected due to a lack of fossils, but also because we tend to focus on the brain, the organ of technical intelligence, and the hands. The strange combination between exclusive bipedalism on the ground and arboricolism replaced hominization in its corporal, carnal, muscular and physiological dimensions. Nevertheless, the brainstem, the spinal cord and the medullary cavity of the vertebrae, that is the segmented body of the embryo all along the dorsal cord, were still missing.

The first year (1982–1983) was dedicated to the *Diplôme d'Études Approfondies* (DEA – Diploma of Advanced Studies) and devoted to the mandible, without preconceived ideas. This bone is the most frequent fossil found in collections, while conversely, intact skull bases are very rare. *Statistical values related to the posture and the angle of straightening are therefore more numerous than for any other bone of the post-cranial skeleton*. The dissertation described the adult mandibles of *Homo sapiens* and the African great apes, with chimpanzees and gorillas classified in the now-obsolete Pongidae family, then it synthesized the anatomical and metric knowledge of the mandibles of Hominidae, with the *Australopithecus* of Afar and *Homo erectus* of North Africa named "Atlanthropus". They were discovered in Tighennif (Ternifine) in Algeria by Camille Arambourg.

The DEA dissertation concluded on the most characteristic part of the differences, namely, the symphysis that unites the two hemi-mandibles in the plane of symmetry. Inclined forward in great apes, it verticalizes in stages for *Australopithecus*, North-African *Homo erectus* and finally in *sapiens*, regarding the chin. The PhD in Human Paleontology (1983–1987) hypothesized the existence of a dynamic link between the symphysis and the spine in opposite directions, and thus with the mode of locomotion. Its confirmation depended on quadrupedalism or bipedalism. Nevertheless, uncertainty persisted since the *Hylobatidae* (gibbon and siamang) are bipedal (Figure 5.1), whereas their symphysis is more prognathic than a great ape's. The locomotor influence therefore seemed contradictory.

Figure 5.1. *Two siamangs in bipedal locomotion holding hands (photo: A. Dambricourt Malassé). For a color version of this figure, see www.iste.co.uk/dambricourt/embryogeny.zip*

The internal organization of the skull base was not mentioned in the articles on the mandible and was unknown to me, as it was for Lamarck 180 years earlier. But the idea of coordinated morphodynamics between the different parts of the same system seemed obvious after a university education in Earth Sciences and professional training in rheology (the study of the behavior of a solid subjected to different stresses: elastic (reversible) or plastic until rupture). Thermodynamics was taught in sedimentary geology with practical works (precipitation of mineral salts, such as calcium carbonate according to pressure and temperature conditions). It seemed normal to study the cranial morphogenesis rich in calcareous salts (calcium phosphate and carbonate) with a dynamic approach, tracing the geometric modeling of its growth vectors.

The hypothesis of a dynamic relationship between the symphysis and the spine was therefore the subject of the thesis (1983–1987), entitled "*Ontogenèses, paléontogenèses et phylogenèse du corps mandibulaire catarhinien. Nouvelle interprétation de la mécanique humanisante* (*théorie de la fœtalisation*, Louis Bolk 1926)"[1]. "Humanizing mechanics" referred to fetalization theory. The anatomical

1 Translation: Ontogeneses, paleontogeneses and phylogenesis of the catarrhine mandibular body. New interpretation of the humanizing mechanics (Louis Bolk 1926).

and metric study of mandibles was completed with radiographies of skulls. A doctor in medicine and radiologist, Jean-François Lepreux, brought Robert Gudin's pantograph to my attention, which already modeled the movement that was posed as a working hypothesis for the thesis. A synthesis of publications on the embryogeny of both the mandible and skull base placed in their cephalic context, in primates and other mammals, revealed that *there were no plicatures at the very beginning of organogenesis*. The symphysis straightens with the straightening of the clivus located below the embryonic nervous system or neural tube. *The fetalization theory was incorrect. The pantograph closed during phylogenesis and no one knew the cause, but what was certain from then on was this: the origin is embryonic and locomotion has never been the living mechanism at the origin of permanent bipedalism.*

Against all expectations, the evolution of the mandibular shape, theoretically confined to manducation and vocalization, was revealed to be closely linked to the complexification of the nervous system from embryogenesis onward, not only in relation to the number of neuroblast divisions, but also with intra- and extra-cellular matrix tensions of the neural tube. Even the dental embryogeny partially induced by the neural crests has changed, especially the canine, the first deciduous molar and the first premolar, which contribute to occluso-postural balance. Straightening is acquired at eight weeks, the end of embryogenesis. Then, during fetal and post-natal development, the functions and specializations differentiate progressively within the limits of the ontogenetic organization plan.

A dynamic unity made possible by the embryonic connective tissue could be observed from the ventral end of the cartilaginous mandible to the caudal end of the embryo, passing through the pre- and basi-sphenoid which were originally aligned.

This is how the physical mechanism at the origin of the angulation of the base, or the rotation of the basi-sphenoid, was highlighted. The rotation is the tip of the dynamic "iceberg" formed by the organogenesis of the neural tube and its tensions (Figure 5.12).

The general structure of Homo sapiens *and the conditions of its future occluso-postural balance are thus set up during organogenesis.*

Thus, on the one hand, axial verticalization during phylogenesis appears to be the consequence of the changes of these tensions and shaping of supporting tissues (bone, cartilage); on the other hand, the different hominin bipedal postures (*Australopithecus*, *Homo*) appear as the consequence of the amplification of the embryonic straightening of the great ape. The distinction between *Pongidae* (*ponginae* and *paninae*) and *Hominidae* is biologically coherent, whereas the current classification of the *paninae* (gorilla and chimpanzee) within *Hominidae* is a radical biological misunderstanding. No embryonic development is common to *Australopithecus* and *paninae*. The ancestor of *Australopithecus* is the embryogeny common to *ponginae* and *paninae*, in other words, to Eurasian and African great apes fossils. *It is not a specialized locomotion of ancestral species that has changed, but the most generalized level of organization, the organogenesis.* Arboricolism simply attests to a way of life in a predominantly arboreal environment, compatible with the psychomotor development of children who were more *internally straightened* at birth, with a mode of movement where they lean exclusively on the two lower limbs, a posture kept until adult stage.

Permanent bipedalism appears as a threshold of increasing complexity of embryogenesis, especially the neurogenesis, the threshold of an internal evolutionary process operating during gametogenesis. The years dedicated to the completion of the thesis thus remain a period of this discovery within the Lamarckian school and more precisely at the IPH. As a famous physicist would later say, the discovery was a Copernican revolution.

This reversal of perspective profoundly transformed the understanding that Homo sapiens *have of their evolutionary identity, here and now. We are the last stage of a process and to wonder about warning signals is a matter of responsible awareness.*

The living conditions are not excluded, quite the contrary, because, as Camille Arambourg pointed out, species evolved together and rapidly when the living conditions changed. *It is still necessary to understand these conditions and how the gametes may have reacted to them.*

Now, the shape, dentition, occlusion and function of the mandible are integrated in the postural equilibrium, and accepted in posturology and dentofacial orthopedics. However, like Gudin's dynamic pantograph, this clinical knowledge is kept hidden in human paleontology, in spite of 30 years of congresses, symposiums, DEA and PhD theses. A Discovery Channel documentary by French producer Thomas Johnson (2005) introduced this situation with a historical debate at the *Institut de France*, held during the international colloquium "Climates-cultures-societies during Prehistoric times, since hominid advent up to the Neolithic", organized by the *Académie des Sciences and the Académie d'Inscriptions et Belles Lettres* (de Lumley 2006; Dambricourt Malassé 2006a). This was not to question Lamarckian doctrine, as such, as it was to question the exclusion, through synthetic theory, of both orthogenesis and punctuated equilibrium (saltationist emergence). Indeed, the curve of increasing complexity/consciousness, brought together by Teilhard de Chardin but limited to the brain, is also confirmed for locomotion, and not by gradual and random mutations, but by threshold effects of increasing complexity during gametogenesis.

The thesis was a novelty, due to the enlightening of internal duration, the thermodynamic arrow of time, with dynamics at the origin of the phylogenetic straightening. And consequently, disciplines little known in paleontology, with the nonlinear dynamic systems and their mathematics, were integrated. In biology, the equations formalize the trajectories as a function of internal duration, that is, the causal chains that have contributed to the organization at an instant t. These sciences have their own vocabulary, such as basin of attraction and attractors, classical in astrophysics as in theoretical biology (Dambricourt Malassé 1992, 1995, 1996a; Bois 2010). The difference between the rotation of the Earth around the Sun, and the biology, is the integration of duration in the cellular construction which contains, one does not know how, its first principle which is the memory, since the finality is the reproduction of the memory of the construction.

> Computing basins of attraction results in a new understanding of the behavior of dynamical systems. The basin of attraction of an attracting set is the set of all the initial conditions. (Nusse *et al.*, *Basins of Attraction*, 1994, pp. 269–314)

And 20 years later:

> Living organisms evolve, in part, according to the underlying properties of the amino acids and other compounds of which they are composed. Thus, there are evolutionary basins of attraction that living organisms will tend to evolve toward. These processes are complex and probably beyond our current capabilities to fully envisage. But progress is being made toward an understanding of such principles by efforts to catalog protein folds and protein–protein interactions. Even plants and animals show convergent evolution, possibly driven by underlying evolutionary basins of attraction. (Gardiner, *Evolutionary basins of attraction and convergence in plants and animals*, 2013)

The science of nonlinear dynamical systems is at the antipodes of the Princeton synthetic theory which cannot integrate it, since this doctrine does not recognize the direction as a consequence of internal logics, such as molecular catalysis capable of amplifying a chain reaction with threshold effects in the long term. The initial conditions cannot be integrated, since only accidental genetic mutations, without causality (i.e. without any relation with the past of the gametes), are conceivable. The notion of memory or memorization is absent, because it constitutes a determinism incompatible with the doctrine of chance as the only conceivable causal agent. The most radical opinion is that of geneticists who have no knowledge of skeletons and fossils, but who claim that no modeling of the straightening for prediction is conceivable, as it would be a "teilhardian aberration" (Langaney 2006), whereas if complexity can self-amplify during gametogenesis, verticality tends toward a physical limit with occluso-postural and psychomotor balances, which can multiply the imbalances.

Moreover, the emergence of reflexive consciousness in self-organized cosmogenetic processes remains a question addressed with the probability of extraterrestrial civilizations by the French CNES (National Center for Space Studies), the French Astrobiology Society (SFE) and the International Academy of Astronautics (IAA) (Dambricourt Malassé 2011a).

The thesis reinforced the relevance of the foundation for a documentation center on Pierre Teilhard de Chardin's works at the *Jardin des Plantes* and PhD theses on the evolution of self-organized systems with mathematics, bio-physics and bio-thermodynamics. The scientific synthesis was recognized, but with great shortcomings in the human sciences, on anatomy, organogenesis and ontogenesis, and above all, its study of the nervous system, its skeleton and psychomotor development, which did not help in our efforts to understand the construction of the symbolic mind and the reflexive consciousness. This foundation was not, therefore, a center solely for the study of his thinkings, since the *Muséum* could not favor a single scholar, but it did serve science and humanism and was justified because of a dogmatic context that did not recognize orthogenesis (or internal determinisms), especially for human beings, of which Teilhard de Chardin was the leader. The *Muséum* thus fulfilled its mission, the fragility of which Paul Broca, and later Henri Breuil, had expected, since at the top of the scale of natural beings, humans remain the ones in search of our natural origins, expressed by equations, and our ontological identity, expressed by symbols (Dambricourt Malassé 2001b).

Since the thesis was defended at the IPH, 30 years of research continued at the *Muséum national d'Histoire naturelle* (1990–2020) in order to confirm the discovery, which did not only concern the mandible, but also the skull base with which it articulated. The constitution of medical images of current and fossil species to complete t statistics was a very long-term program. The rarity of fossil skull bases, the inefficiency of X-rays too opaque when the bones were mineralized, the distance of the fossils preserved in their countries and finally the difficulty of accessing the very first CT scans in France, explain the great slowness in setting up such a database. The context can be compared to the working conditions of Alfred Velpeau, who took years to build up a usable collection of human embryos. *It is therefore pioneering research that had at its disposal a large number of collections from the* Muséum, *the* Muséum de l'Homme *and the* IPH.

With rare exceptions (Hadjouis 1999, 2018, 2021), paleoanthropology does not show interest in advances in posturology and occlusodontics that place the sphenoidal synchondroses and their dynamics at the center of correlations. The mandible of current and fossil primates is therefore not yet studied according to the ontogenetic approach, placed in its basicranial and endocranial context. Gudin's pantograph and craniofacial biodynamics remain the guiding thread of research programs on postural balance and the issues at stake in the understanding of hominizing processes and their becoming.

Emblematic fossils and constructive controversy

Paleontological collections have been enriched and several fossils now allow a better understanding of modalities of encephalization and straightening processes. The lag of human paleontology, compared to the advancement of posturology and occlusodontics, can be observed with the interpretation of the fossils from the Moroccan site of Jebel Irhoud, 300,000 years old and attributed to the earliest *sapiens* (Hublin *et al.* 2017). The sphenoid bone is absent, the diagnosis is established on the shape of the vault, brain volume and mandibles, but the position of the occipital squamous and the inclination of the petrous pyramids are sufficient to note a sphenoidal dynamic closer to a Neanderthal than to a Rhodesian Man (200,000 years old) and unrelated to *Homo sapiens*. Consistently, the mandibles do not have a chin. The positions of the brain stem and the spinal cord are first in embryogenesis; they precede brain growth. The Jebel Irhoud skulls do not have the more pronounced straightening that distinguishes *Homo sapiens* from any fossil species as soon as there is an onset of embryogeny. These fossils testify to different complexification processes of the cerebral hemispheres, but not to the emergence of *sapiens*, which concerns the whole nervous system since embryogeny. Understanding the emergence of *sapiens*, whether in Africa, Asia or in a fertile crescent between East Africa and South Asia, cannot progress without embryogeny and craniofacial biodynamics; Gudin's pantograph and the kinetics of the neural tube are unavoidable.

The hominization curve became exponential between the first hominids – or the first verticality threshold (4 million years ago) – and the last threshold crossed around 200,000 years ago – *Homo sapiens*. This is five times faster than between the first semi-erect great apes (20 million years ago) and the first hominids. The probability that this production of complexity will continue cannot be ignored, we do not know the mechanisms but we can anticipate the effects. As the French mathematician and Fields Medal winner, René Thom (1923–2002), said: "To predict is not to explain". The naturalists who have always shown an interest in this discovery are the practitioners encountered in osteopathy and orthopedics, because disorders of posture, cerebellum development and occlusion are perhaps symptoms. Therefore, the scientific attitude is one of the fundamental research with geometrical formalizations and statistics that can be integrated by the sciences of nonlinear dynamic systems and their modeling for predictions.

5.2. Evidence of craniofacial contraction

5.2.1. *Materials, method and objectives*

The acquisition of the metric and anatomical data took place in two phases, the first during the thesis years, and the second during the 30 years of research at the IPH, headquarters of the Prehistory laboratory of the *Muséum national d'Histoire naturelle*.

The first phase was the study of the mandibular body in its cranial context (1983–1987). The material consisted of mandibles and skulls from the collections of the IPH from all continents, and from the *Muséum national d'Histoire naturelle* kept in the Comparative Anatomy Laboratory, the Paleontology Laboratory and the *Musée de l'Homme* (Table 5.1). The objective was to model the growth and morphogenesis of the body of the mandible in the sagittal and transverse planes, placed in the context of cranial morphogenesis and growth. Morphogenesis was described in all three planes, but skull growth trajectories were only geometrized in the sagittal plane (in profile) by a combination of both Delattre and Fenart, as well as Gudin's traces. The orthonormal landmark was the Frankfurt plane with the porion as origin (point above the external auditory meatus). For the vault, the craniometric points were the bregma, lambda and inion. Measurements of the mandible were simple, using a caliper, a goniometer and a dioptograph to trace cranial profiles. However, as the measurement of the sphenoidal angle was only possible on radiography, the acquisition of statistics required the setting up of a large-scale program. Such a project could only be undertaken by the management of a laboratory with agreements between a hospital teleradiography[2] center and the curator of the collections. A first limited series of radiographies was produced to illustrate the values of the sphenoidal angle of the primates that have already been published by Gudin and Delattre.

The method was therefore the definition of linear and angular measurement protocols, applied to the largest number of mandibles, from the fetal stages preserved in the collections of the different laboratories of the *Muséum*, Comparative Anatomy, Paleontology, the IPH and the *Musée de l'Homme*. The measurements in the two sagittal and transverse planes allowed the calculation of statistics by dental age class, for different fossil and current taxa of the great groups of Catarrhine, classified as monkeys, great apes and hominids, that is the Australopithecine and the fossil species of the genus *Homo*. For the current species, the material totaled 507 mandibles and 25 radiographies of the skull in profile including mandibles, carried out by the Comparative Anatomy Laboratory and with

2 Teleradiography is a radiography that uses an X-ray tube placed 2 meters from the head to obtain a full-size image of the skull.

Dr. Jean-François Lepreux at the Medical Imaging Center of the La Montagne Clinic (Courbevoie). The fossil collection consisted of original mandibles and casts, for a total of 117 mandibular bodies. The current *Homo sapiens* (n = 208) and fossil (n = 115) collection belong to the IPH, except for the mandibles of very young fetuses, which belong to the *Musée de l'Homme*. The fossil mandibles of monkeys and hominids are originals and casts. For monkey fossils (n = 23), the original mandibles come from the Paleontology collections, and the casts come from the IPH. The mandibles of *Australopithecus* and *Paranthropus* (n = 35) are the casts from the "International Afar Research Expedition" of Maurice Taieb, preserved at the *Musée de l'Homme*, and those from the IPH. The non-*sapiens* fossil species (n = 30) are original mandibles and casts from the IPH, made just after the discovery of the fossils. Concerning the skulls of living species, all monkeys, apes and *Homo sapiens* were original. For profile contours, fossils of *Australopithecus africanus* and *Paranthropus* and of non-*sapiens* species (n = 16) were casts from the IPH collections drawn with a pantrograph.

The second phase was the modeling and statistical measurement of craniofacial contraction (1988–2020). Although the study of fossils by tomography (CT scanner) is now systematic, this was not the case for a 20-year period. The imaging database was first enriched with radiographies because medical scanners were non-existent and then very rare. The skulls could be scanned thanks to several collaborations with radiologists.

– *Teleradiographies of skulls*:

The radiographic database was composed of skulls from the collection of the IPH in collaboration with Dr. Florence Vuylsteker, dental surgeon, and the department of Professor Jean-Claude Libersa of the Regional Hospital Center of Lille, with a total of 318 teleradiographies in both sagittal and transverse planes, according to the standard norms of Dentofacial Orthopedics with *Homo sapiens*: 110 adults, of all geographical origins; *paninae*: 25 skulls including 10 *Gorilla* and 15 *Pan* (13 adults and 2 youths); and *Cercopithecidae* (*Macaca)*: 24 with 10 youths and 14 adults.

– *CT scans of current skulls and fossils*:

Collaborations:

1) With Dr. Fabienne Lallouet, radiologist for dry skulls of human fetuses, including an exceptional anencephalon without basi-cranial flexion from the collections of the *Musée de l'Homme*, scanned at the *Groupe Hospitalier Diaconnesse-Croix Saint-Simon* in Paris. In addition to these fetuses, CT scans of anonymized patients from the Picture Archiving and Communication System (PACS) of the American Hospital in Paris were performed on 55 individuals,

including 30 adolescents aged 12–17 years and 25 adults aged 23–64 years, in equal numbers of women and men. To *Homo sapiens*, skulls of paninae and ponginae were added from the IPH collections, with a five-month-old gorilla fetus and 41 skulls of *Pan troglodytes* chimpanzees divided into 32 youngsters from three months to nine years old, and nine adults, as well as three orangutan skulls: one adolescent, one female and one male.

2) With Prof. Eric de Kerviler and Dr. Jean-Pascal Martin of the Teleradiography Service of the Saint-Louis Hospital in Paris: skulls from the IPH collections with *Homo sapiens* fossils: 95,000 year-old from Qafzeh (Israel), including Qafzeh 6, Qafzeh 7 and the mandible Qafzeh 4, and six young chimpanzees.

– *Muséum* collections and research programs:

The total number of scans was 57 current *Homo sapiens*, 51 great apes (chimpanzee, gorilla and orangutan), nine Australopithecines, three *Homo erectus* from Java, four Neanderthals from France (three adults, one adolescent and one child), three 95,000-year-old *Homo sapiens* and two 27,000-year-old *Homo sapiens*. These scans come from:

1) The fossil collections of the *Musée de l'Homme*: *Homo sapiens* with the complete skull of Cro-Magnon 1 and the Abri Pataud and four Neanderthals with La Chapelle-aux-Saints, La Ferrassie 1, La Quina (teenager) and the Pech de l'Azé (child).

2) The research program "*Hominids in the Quaternary: Environments and Behaviors*" in Java, directed by François Sémah, professor at the Prehistory Department of the *Muséum national d'Histoire naturelle*, who has entrusted three *Homo erectus*, Ngandong 7, Ngandong 12 and Ngawi 1.

3) The collections of the Witwatersrand (South Africa). The CT scans of *Australopithecus* and *Paranthropus* (Australopithecines) were entrusted by Phillip Tobias (1925–2012), the inventor of the taxon *Homo habilis*, with John Napier (1917–1987) and Louis Leakey (1903–1972):

– the skull of Taung;

– Sterkfontein Karst: skulls StW 505 and Sts 5 and mandible StW 498;

– Swartkrans Karst: skull SK 48;

– Makapansgat Cave: skull MLD 37/38, mandible MLD 2 and maxilla MLD 9 and MLD 40.

5.2.2. *Protocol and results for current species*

5.2.2.1. *The mandible*

In the sagittal plane, the protocol measures the symphyseal angle formed between the most anterior point of the basilar edge of the mandibular body (*pars basilaris*) – or gnathion – and the bony point between the two incisor alveoli (*pars alveolaris*). In the transverse plane, *a new angle is defined at the gnathion: the basilar angular deviation between the two hemi-mandibles*. The angle is calculated by the sinus given by two linear measurements: on the one hand, the length L of the mandibular base, measured from the gnathion to the optimal inflection point of the pre-angular notch, and on the other hand, the distance D between the two notches: sin α = ½ D/ L (Figure 5.2(A)).

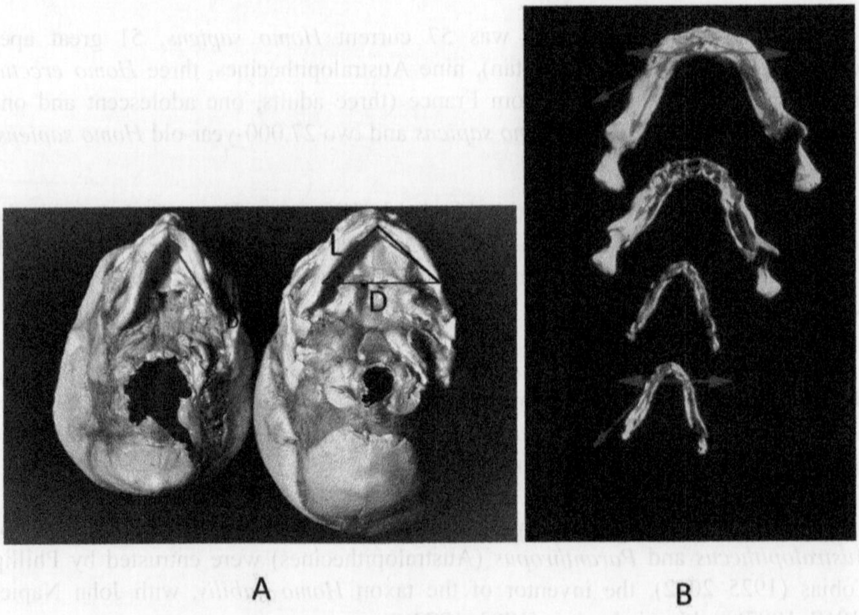

Figure 5.2. *A: Inferior view of human fetuses at the same scale, aligned with the spheno-occipital synchondrosis (horizontal line) showing the spacing between the two hemi-mandibles due to rapid transverse brain growth. L and D are explained in the text. B: Fetal mandibles in occlusal view showing the frontalization of the incisor-canine alveoli as the two hemi-mandibles move apart from each other (photos: A. Dambricourt Malassé). For a color version of this figure, see www.iste.co.uk/ dambricourt/ embryogeny.zip*

Linear measurements are the height of the symphysis and the length of the *pars basilaris* between the gnathion and the inflection point. The shape of the alveolar arch was separated into two segments, the first to quantify the incisor-canine frontalization, and the second to measure the divergence of the molar alveoli from the incisor-canine segment (Figure 5.2(B)). Alveolar and dental arch angles of the mandible had just been measured by Helga Roth (1983) on Australopithecines, all non-*sapiens* fossil taxa of the genus *Homo*, and fossils of *H. sapiens*. These values were sufficient to calculate the angles of alveolar frontalization and divergence.

5.2.2.1.1. Homo sapiens

The fetal growth of *Homo sapiens* was measured with 35 specimens from the third intrauterine month; split into 13 fetuses of increasing age and 23 from neonatal stages. Post-natal growth was divided into four phases according to the dentition: the growth rate of the dimensions and angles between the phases allowed us to follow the differences in speed in the two planes, sagittal and transverse (Table 5.1).

Phase	Dental phase	n
Phase 0	Newborn and absence of teeth	23
Phase 1	Deciduous teeth	17
Phase 2	Eruption from M1 to M1 in place	20
Phase 3	Eruption from M2 to M2 in place	16
Phase 4	M3 in place, adult	119

Table 5.1. *Number of specimens studied by phase of dentition*

What is remarkable is that the opening of the basilar angular deviation is from 60° to 89° ± 8. The angle reaches 70° in the fourth month, then remains at 82° until the eighth month, and then 90° is quickly reached during the ninth month. This opening corresponds to the lateral growth of the skull base between the two temporo-mandibular joints, related to the expansion of the temporal lobes of the brain. The shape of the mandible is therefore initially long and very narrow. The alveolar arch shows a closed frontalization angle, it opens as the two hemi-mandibles move away from each other, driven by the articular condyles. Figure 5.2(A) illustrates the rapid lateral growth of the temporal lobes above the condyles, because the length of the mandible has not changed. These are true displacements.

The bony trabeculae structure of the hemi-mandible shows an organization in the alveolar part (*pars alveolaris*) in the shape of a basket (see Figure 5.3, left). The trabeculae of the basilar part (*pars basilaris*) extend backward to the condyle that articulates the mandible with the skull base, while the two anterior ends in contact in the sagittal plane move away from each other, leaving a slightly outlined triangular vacuum on the three-month-old fetus. The bony trabeculaes of the alveolar basket of both hemi-mandibles go up in the anterior part and obliquely move toward the plane of symmetry to meet. The divergence of the trajectories between the rectilinear basilar part and the *pars alveolaris* forms a depression, named the "chin fossa" (*fossae mentales*).

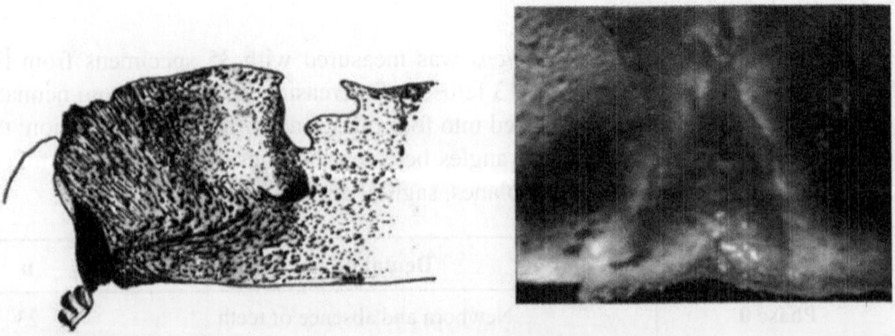

Figure 5.3. *Left: the symphyseal vacuum (black triangular space) in the human neonate with the orientation of the bony trabeculae, upward for the alveolar basket, and straight forward for the basilar part. A depression is formed between the two sets of trabeculae. Right: a very young fetus with the symphyseal vacuum in the process of ossification bordered by the chin fossa (drawing and photo: A. Dambricourt Malassé (1987)). For a color version of this figure, see www.iste.co.uk/dambricourt/embryogeny.zip*

The symphysis angle is 90° until the end of the fetal period. After birth, the symphyseal space becomes ossified and its edges disappear, the chin clears variably according to the recoil of the alveolar arch and the symphyseal angle tends to close with an average value of 83°± 10.

The optimum of linear growth in the three planes of space corresponds to the setting of the temporary dentition, but the lateral growth is slower than the longitudinal, so the basilar angular deviation closes from 90° to 77°± 6 in the adult stage. The shape of the alveolar arch becomes parabolic. This relative closure means that the growth of the face eventually predominates over that in the transverse direction of the brain. In the inferior view, the basilar part of the chin shows on each side of the symmetry plane, a flat surface corresponding to the insertion of the digastric muscles that support the hyoid bone under the tongue. All of these details are important because they will allow us to distinguish the symphyseal region of *Homo sapiens* among all the fossil mandibles, as already predicted by the critical opinion on the fragment found in Kanam by Louis Leakey in 1932.

5.2.2.1.2. Great apes (paninae, ponginae), gibbon (Hylobates) and *Cercopithecus*

Table 5.2 shows the number of individuals by dental phase and genus.

Phase	*Pan*	*Gorilla*	*Pongo*	*Hylobates*	*Cercopithecus*
Phase 0	0	2*	0	1	1
Phase 1	2	9	1	6	1
Phase 2	25	6	11	3	18
Phase 3	17	4	0	6	11
Phase 4	35	45	16	16	30

Table 5.2. *Number of specimens studied by dental phase. * one fetus and one neonate*

Paninae and ponginae

The mandible of the three genera, *Pan* and *Gorilla* (paninae) and *Pongo* or orangutan (ponginae) has the same conformation at the youngest stages in temporary denture, visible at the five-month fetal stage (one specimen) and remain until the adult stage. This conformation has nothing comparable to *Homo sapiens* (Figure 5.4).

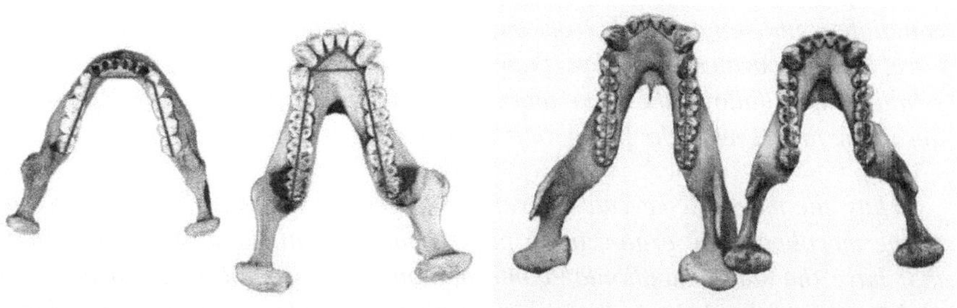

Figure 5.4. *Comparison of adult dental arch and mandibular body shapes. Left:* Homo sapiens *compared to gorilla. Right: gorilla and orangutan. The mandibles are at the same scale (photos: A. Dambricourt Malassé). For a color version of this figure, see www.iste.co.uk/dambricourt/embryogeny.zip*

The symphysis is sagittal over the entire height and inclined forward, the two basilar edges meet forming a small plate called a "simian", surmounted by a dimple, the genio-glossal fossa, into which the lower muscle of the tongue is inserted. This orifice is dominated by the *alveolar planum* which slopes gently up to the alveolar edge. The upper view shows this conformation and the difference with *Homo sapiens* (Figure 5.4). The alveolar arch is U-shaped, known as upsiloid, with a much less pronounced incisor-canine frontalization than *Homo sapiens* and the molar alveoli parallel. However, the basilar angular deviation is 90° at birth and therefore identical to *Homo sapiens.* In other words, the basilar part of the mandible, which extends up to the condyle, followed the same fetal transverse growth in *Homo sapiens* and the great apes. On the other hand, the alveolar part (*pars alveolaris*) is uncoupled from the basilar part (*pars basilaris*), because unlike *Homo sapiens*, it did not follow the base in its growth on either side of the plane of symmetry. With *Homo sapiens*, the alveolar part is integrated into the transverse growth of the base. The symphyseal angle is 120° ± 2 at birth in all three genera. Then it opens most rapidly up to 136° ± 3 in the male gorilla, while in the chimpanzee and the orangutan the average value is the same with 132° ± 5. The basilar angular declination closes up to 64° ± 7 in the male gorilla, with a mean of 71° ± 9; then 74° ± 6 in the chimpanzee and 75° ± 8 in the orangutan.

At birth, the basilar part is therefore wide, but the U-shaped alveolar arcade does not follow its contour and the forward sloping symphyseal region does not form a symphyseal vacuum; chin fossa are physically impossible. The forward alveolar growth of the gorilla is the most pronounced, relative to the transverse plane, since the closure of the basilar angular deviation is maximal, and relative to the basilar part, since the symphyseal angle opens. The adult male gorilla accentuates the growth trends of the females and the other two genera.

The first observation is the morphogenetic unity of the three genera. A gorilla's mandible is indistinguishable from that of the orangutan, even though their lineages have been separated for more than 14 million years. That of a chimpanzee is smaller, but it follows the same morphogenesis. The difference with Homo sapiens *lies at the first level of the living, that of the ontogenetic organization plan.*

Thus, the mandible reveals a level of ontogenesis organization that corresponds to the morphogenetic organization plan common to all fossil and current species that have the same mandibular conformation, whatever the stage of development and the size of the skull. And it is this morphogenetic plan of organization that is the ancestor of Homo sapiens.

This biological unit, the first of all, has no name because it has never been highlighted; it does not correspond to the Linnaean nomenclature, it is neither a species, a genus, a family nor an order. I have chosen a neologism, "fundamental ontogeny", to define this first level of organization of the animal which begins with embryogenesis (Dambricourt Malassé 1987, 1988, 1992, 1995; Saban 1993; Dambricourt *et al.* 1999) and was published in the *Encyclopedia Universalis* in 1999 (Chaline 1999). In the absence of a terminology consistent with embryogenesis and phylogenesis, the word "great ape" has been retained. It excludes *Homo sapiens* and all small monkeys, as well as *Hylobatidae* (gibbon, siamang).

The most notable difference is the disjunction of the growth trajectories of the alveolar arch – *pars alveolaris* – from the basilar part – *pars basilaris* – whereas in *Homo sapiens*, it follows the lateral trajectories of the base as early as fetal growth. Since the alveolar parts of the mandible and maxilla overlap, these neonatal differences correspond to a different morphogenesis of the whole face from the fetal period. It is therefore necessary to compare the differences between the maxillary arch of the human fetus and that of the great ape morphotype, in order to identify the difference during development that allows humans to form a single whole, both at the level of two alveolar arches, and at the level of the posterior base of the skull where the mandible articulates with its condyle, following the straightening movement of the skull base. These few observations are already sufficient to predict the location of the differences, the zone where the modifications must be coordinated between the spinal field and the facial field already observed by Deshayes' cephalometric analysis (Figure 4.5).

This area of coordination between the face and the spine is necessarily in the center of the base of the skull, in other words, somewhere in one of the synchondroses of the sphenoid. These differences between the trajectories of the "great ape" morphogenesis and that of *Homo sapiens* are reminiscent of Weidenreich's transverse axis of rotation, around which the skull wraps with retrognathism. Spheno-occipital synchondrosis is often seen as the primary center of base rotation, but Björk and Scott have made it a secondary center of angulation named spheno-basion by Beauvieux (Olivier 1965, p. 88). We will see in the section on embryogenesis that the first axis of rotation activated is indeed the synchondrosis between the pre- and post-sphenoid.

These morphological changes correspond to dental anatomical and occlusal changes for temporary and permanent teeth. The dental formula (or the number and type of teeth) is the same in great apes and in *Homo sapiens*: two incisors, one canine, two milk molars replaced by two premolars and three permanent molars. But notable differences from the milk teeth, or temporary teeth, are observed with the occlusion between the canines. In apes, their crowns are higher than those of the incisors and milky molars, and their position is reversed in relation to *Homo sapiens*'. The lower canine is placed in front of the upper canine that creates a space behind the maxillary incisor to house its apex, the diastema (Figure 5.5). The lower canine is blocked in this maxillary space. The upper canine is wedged between the lower canine and the lower first molar. This lower molar is different from the others because its occlusal surface has only two reliefs, or cusps, instead of five. It is said to be "caniniform" because the cusp near the canine is protruding and pointed, with a wear facet characteristic of friction with the upper fang. British schools retain this wear as a chewing specialization (White *et al.* 2015), and name this singularity C/P3 honing complex (P3 being the first premolar). The smaller the facet of a fossil species, the more limited this sharpening function was, suggesting a proximity to the *Hominidae*, because the latter have no fangs. The hominin occlusion would be linked to bipedalism. However, this C/P3 complex also exists in the gibbon, which is bipedal on the ground.

In *Homo sapiens*, everything is different: the crowns do not protrude beyond the occlusal surface of the other teeth, the incisor-canine dental arch is positioned behind its maxillary counterpart and the lower canine is behind the upper canine (Figure 5.5). Next, all of the mandibular molars have five cusps. The teaching we provide in the *Muséum national d'Histoire naturelle* does not consider a sharpening function, *but an occluso-postural function, called "canine lock", linked to the degree of the axial skeleton straightening; what is significant is not the C/P3 complex, but the C/C complex.* Hominian occlusion would be linked to bipedalism. However, the gibbon also develops the C/P3 and C/C complexes.

The occlusion of the young ape is thus locked by the canines in the anteroposterior direction and laterally. In the adult stage, the difference is magnified as the crowns accentuate the characteristics of the temporary dentition, whereas such an occlusion is missing in *Homo sapiens*, even in promandibulia.

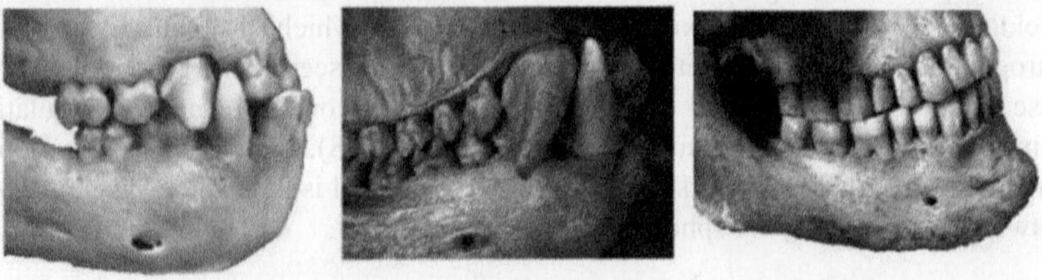

A B C

Figure 5.5. *Incisor-canine occlusion (different scales). A and B, gorilla: canine lock in temporary and adult dentitions. C:* Homo sapiens*: no lock in adult dentition (photos: A. Dambricourt Malassé)*

The canine lock is thus reinforced with the growth of the dental arch forward, in other words with the formation of a sort of muzzle, which lengthens the face and moves it away from the brain cavity. We will see that these differences in occlusion are correlated with the anti-clockwise direction of the sphenoidal angle which opens during growth (the face on the left). The mandibular condyles are drawn into this trajectory and there is no retromandibular position thanks to this locking. The canine lock will be all the more pronounced as the occipital territories are solicited by tensions in quadrupedal mode, or even by tensions of the shoulder girdle (scapula and clavicle) in arboreal mode. The wear facet is an effect of the C/C complex, the less-pronounced canine lock in a great ape fossil species may be a generalized and interesting character to include in the ancestral lineage of the hominids. Nevertheless, the conceptual difficulty, which was at the heart of the debate of the colloquium of Paris in 2004, remains the transition that was not gradual and functional, but saltationist and embryonic, as allowed by homeotic genes regulation (Dambricourt Malassé 2006).

The gibbon (*Hylobatidae*)

The skulls and mandibles come from the Comparative Anatomy collections. The size is much smaller than current great apes and *Homo sapiens*, the adult symphysis measuring 2 cm in comparison to 3.5 cm in *Homo sapiens*, 4.5 cm in chimpanzees, 6 cm in orangutans, 6.5 cm in male gorillas and 5 cm in female gorillas. Morphogenesis and fetal growth do not overlap with that of a great ape. At birth, the angular deviation is the same as in the earliest fetal stage of *Homo sapiens* (three months) with 60° ± 6, and the angle closes to 52° ± 4 in the adult stage. The symphysis angle is stable from 125° ± 4 to 123° ± 5.

The temporary and adult teeth have the same formula as the great apes and *Homo sapiens*. The molars and canine follow the same morphological pattern as the great ape with the canine lock, and the crowns are longer and more tapered. The general shape of the alveolar arch is upsiloid, but much more narrow, and the incisor-canine frontalization is less pronounced. The most significant is the absence of basilar angular deviation during fetal growth. This angular value is therefore highly representative of a complexification of the fetal cerebral growth in great apes compared to *Hylobatidae*.

The *Cercopithecus* (*Cercopithecidae*)

The skulls and mandibles come from the same collections of Comparative Anatomy with a total of 60 specimens including 30 adults. The mandible is larger than that of the gibbon; at the adult stage the symphysis measures 2.6 cm on average. The symphyseal angle opens clearly from 123° to 137° ± 4, the angular deviation is 60° at birth and it closes up to 54° ± 3. The dental formula is the same as that of the gibbon, great apes and *Homo sapiens*. The canine lock is ensured by large upper and lower canines. The shape of the mandible is upsiloid, narrow, with an incisor-canine frontalization all the less pronounced the more the symphyseal angle is open, or the symphysis recedes forward. The lower molars only have four cusps; they do not have the "dryopithecine" pattern of the five cusps.

5.2.3. *Comparison of mandibular growth trajectories*

The theory of fetalization is invalidated as the human mandible never passes through a great ape fetal stage, especially the symphysis. The anatomy and the growth trajectories of *Homo sapiens*, great apes, the gibbon and *Cercopithecus* match the classification that has always prevailed since anatomists have been comparing primates, distinguishing between monkeys, great apes and humans, the gibbon being qualified as a "small great ape". In fact, the development of the mandible is a characteristic of the *Cercopithecus*, and so they cannot be placed amongst the great apes; *Hylobatidae* are classified as a family of small Catarhines. The dentition allows us to attribute a common ancestor with the *paninae* and *ponginae* since the lower molars have the same pattern, called "dryopithecine". However, it does not allow us to consider it as a great ape. The comparison of mandibular trajectories shows that there is no *Hominoidea* superfamily that includes the gibbon, *paninae* and *ponginae*, and *Hominidae*, because it has no biological reality, so there is no single morphogenetic plan. As seen in Volume 1, the nesting classification is the result of fixism. The inconceivable emergence of a new organizational plan from the old one forces us to now give it another name. The question does not arise for *Cercopithecus* and the gibbon, as they derive from a stem population of small Catarrhines and remain within the limits of this morphogenetic plan. Their lineage has diverged well before the emergence of the great apes' morphogenesis. The fossil mandibles of their lineages allow us to date this divergence. It is now necessary to identify the degree to which the clivus and brainstem are straightened or the cerebellar fossa is lowered in these four groups.

5.2.3.1. *The mandible in its cranial context*

The mandible was placed in its cranial context in order to compare the growth of its angles with that of the sphenoidal angle and the direction of the trajectory of the vault. The skull is oriented in the Frankfurt plane and centered on the porion. The trajectories of the gorilla are described from a five-month-old fetus, with a newborn and a series of increasing ages to the adult stage. Other taxa are studied from birth. In the fetal stage, the bony squamous are so thin that no difference between the outer and inner tables is apparent. The diploe, or spongy tissue that separates them, thickens with age. The internal table serves as an insertion for the dura mater that protects the brain and follows its intra-uterine growth trajectories. The exocranial table will dissociate from it in non-human primates after birth and follow counterclockwise vectors, as Delattre and Fenart have already shown. New exocranial stresses are generated by insertions of the muscular aponeuroses and their function, with the temporal muscle supporting and activating the mandible during mastication, and those of the neck muscles connecting the back of the skull (occiput) to the postcranial skeleton.

Delattre and Fenart also took into account the bone growth factors acting on the outer table such as the somatotropic hormones. The craniometric points of the outer table, which cover the frontal lobes up to the cerebellum in the median sagittal plane, are the bregma, lambda and inion (Figure 5.6). When the face is oriented to the left, in *Homo sapiens* the points follow a clockwise rotation, from front to back for the vault and from top to bottom and forward for the occiput. The largest amplitude is that of the inion, which corresponds to the post-natal growth of the cerebellum. For all other primates, the direction is the opposite, the inion and the lambda move upward and the bregma moves forward with the growth of the face which becomes detached from the neurocranium. The rotation is said to be positive when the trajectory is oriented toward verticality, and negative when it is reversed. Delattre and Fenart have observed the post-natal dissociation of the two cranial tables in non-human primates. The endinion (inner inion), located at the crossing of the cerebral falx and the tent of the cerebellum, does not follow the trajectory of the outer table – that was also well observed in Delattre and Fénart (1960). The most spectacular disassociation is that of the male gorilla, with the formation of crests by the outer table for insertions of the temporal and occiput muscles.

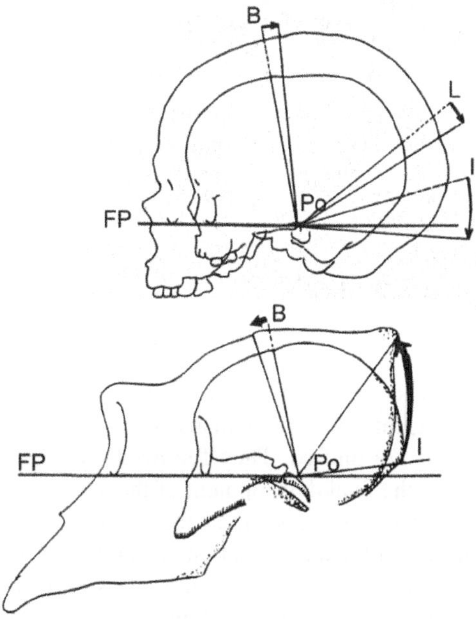

Figure 5.6. *Orientation of the neurocranium growth vectors (A) in* Homo sapiens *toward vertical and (B) inversion of vectors in the great ape (gorilla) with the facial and the skull base growth toward horizontal. The adult inion is on a bony crest. B: bregma, L: lambda, I: inion, FP: Frankfurt plane (Dambricourt Malassé 1987, 1993)*

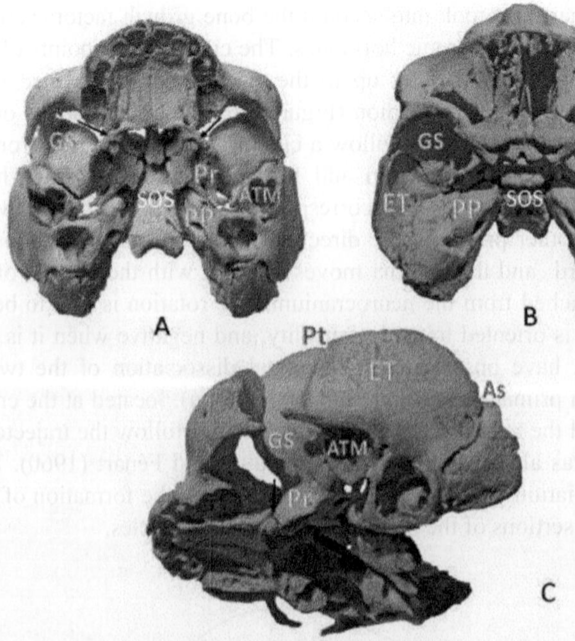

Figure 5.7. *Skull of newborn* Homo sapiens *(incomplete). A: inferior view (exocranium), B: superior view on the endocranium (vault removed), and C: exocranial view in perspective. As: asterion, ATM: TMJ or temporo-mandibular joint, ET: temporal squamous, GS: greater wing of the sphenoid, PP: petrous pyramid, Pr: pterygoid process, Pt: pterion, SOS: spheno-occipital synchondrosis, black arrow: contact between the palatine plate and the pterygoid process (photos: A. Dambricourt Malassé). For a color version of this figure, see www.iste.co.uk/dambricourt/embryogeny.zip*

The innovation is the comparison of the fetal and circumnatal trajectories of *Homo sapiens* and the great apes. In humans, the transverse trajectories of the alveolar and basilar parts are unified. They followed the transverse growth of the middle cranial fossa, thus the fetal development of the temporal lobe. The transverse trajectory of the middle fossa is the same at the level of the temporo-mandibular joints (TMJ), since the basilar angular deviation is identical at birth with a 90° angle. However, it is different in the anterior part and more significant in *Homo sapiens*. Indeed, the alveolar arch of the mandible followed the maxillary arch widened by the palate, which is not the case for the great apes. The palate is attached to the anterior part of the middle fossa by its horizontal plate (black arrow in Figure 5.7(A) and 5.7(C)). This is joined to the vertical pterygoid process (Pr), which is an extension of the greater wing of the sphenoid (GS). The space between the two greater wings has expanded with the lateral expansion of the anterior end of the temporal lobes, as shown by the distances L1 and L2 between the symmetrical sutural points, respectively pterion (Pt) and asterion (As) (Figures 5.7(B) and 5.7 bis).

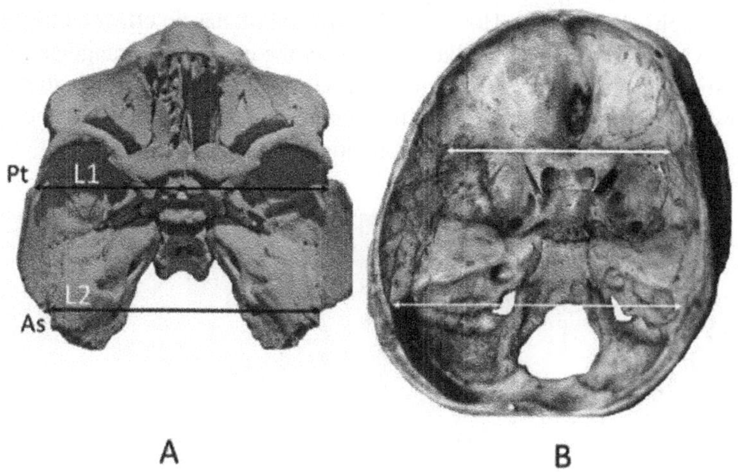

Figure 5.7 bis. *Endocranial skull base of newborn* Homo sapiens *(incomplete) and a young gorilla; L1 distance between the pterions (Pt); L2 distance between the asterions (As) (photos: A. Dambricourt Malassé)*

The neonatal alveolar arch of Homo sapiens *thus shows a greater transverse spread than in the great ape, which corresponds to a more developed lateral growth of the anterior end of the temporal lobes.*

After birth, the psychomotor development of *Homo sapiens* is the slowest of all primates, with incommensurably more axons and synapses and longer myelination. This difference of brain expansion between current great apes and Modern Man is still being explored (Benito-Kwiecinski *et al.* 2021). *The cerebellum develops downward, not driven by the brain, but because the brainstem is more upright from birth and allows the cerebellum to develop downward and forward, this is one of the major innovations of our PhD thesis.* Young apes have no difficulty moving in bipedal mode, as can be observed in the menagerie of the *Jardin des Plantes*, but with age, quadrupedal balance will be even more energy-efficient as the cranial base loses its plasticity following the ossification of the synchondroses and the sphenoidal angle opens.

The ponginae-paninae divergence

In the infantile stage, the chimpanzee's face is almost identical to the gorilla, while the orangutan differs by an upward tilt of the entire facial block, as shown in the radiographs (Figure 5.8). However, this upward tilt has no effect on the mandible and the alveolar arch of the maxilla. This is why the gorilla and orangutan mandibles are indistinguishable. The rotation of the facial block is therefore late in relation to the growth trajectories common with the paninea. The growth zone of the base that allows this facial motion is the spheno-ethmoidal synchondrosis.

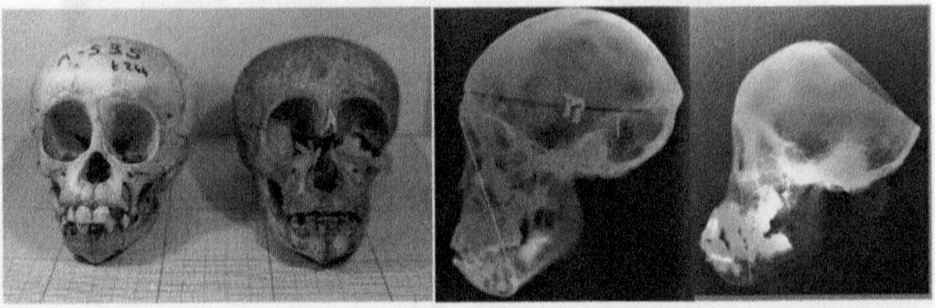

A B

Figure 5.8. *A: Young chimpanzee and gorilla in deciduous dentition, B: radiographs of a young gorilla and an orangutan in mixed dentition. The horizontal line is the reference frame in dento-facial orthodontics and the orientation of the face is given by the major axis of the orbits, passing through the gnathion (photos: A. Dambricourt Malassé (1987))*

As shown by the X-rays, this phenomenon increases with growth and prevents the formation of the supraorbital torus observed in chimpanzees and gorillas, which results from the migration of the exocranial table of the frontal bone caused by the forward growth of the face. Thus, there is a lack of pneumatization of the sinuses which communicate from the body of the sphenoid to the supraorbital arch (large cavities formed between the cranial outer and inner tables).

Delattre and Fenart have already noted the facial motion which explains the conservation of the curved shape of the outer frontal squamous, but they did not give an explanation (Figure 5.9). Deshayes' cephalometric analysis and the consideration of sphenoidal synchondroses shed light on this biodynamic divergence between the cranio-spinal (spheno-occipital) field, which is common to the paninae and ponginae, and the cranio-facial (spheno-ethmoidal) field, which develops later.

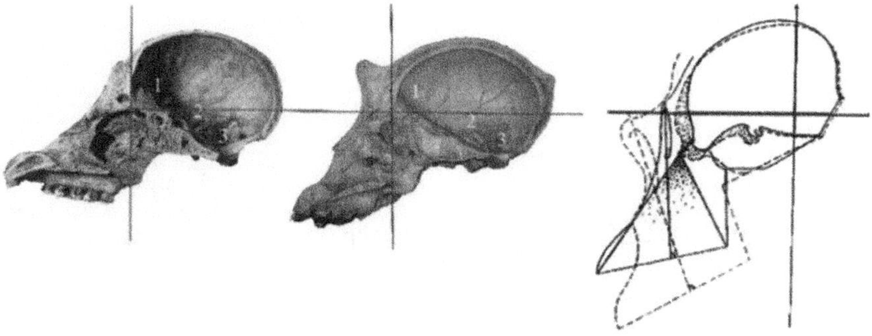

Figure 5.9. *Comparison between adult orangutan and gorilla aligned on the spheno-ethmoidal plane, a perpendicular passing through the most anterior point of the base (foramen caecum), showing the tilt of the face as a block (photos: A. Dambricourt Malassé). On the right, the tilt of the face and the blocking for the supraorbital torus formation oriented in Delatte and Fenart's plane (1960). For a color version of this figure, see www.iste.co.uk/dambricourt/embryogeny.zip*

5.2.3.2. *Collaboration with M.J. Deshayes (1988–1997)*

A collaboration was established with Dr. M.J. Deshayes from 1988 to 1997 to apply cranio-palatal balance analysis to the growth of great apes (Comparative Anatomy collections) and to compare the skull of Cro-Magnon 1 and La Chapelle-aux-Saints (*Muséum de l'Homme*). Agreements between the laboratories and Dr. Philippe Katz (Paris) have allowed the removal of 38 chimpanzees, 26 gorillas and 7 orangutans from the Muséum, for a total of 71 skulls of great apes at different dental stages, and two fossil hominids.

The comparison brought to light a biodynamic discontinuity between *Homo sapiens* and the great ape: whatever the dental stage, the occlusion of the ape is always in class I. Great apes have no cranio-palatal balance from birth (Figure 5.10). The craniofacial (CF) and cranio-spine (CR) fields are equivalent, but the cranio-palatal axis (CP) is not parallel to the sphenoidal plane (C3), whereas it is forward and downward in the human fetus. In apes, after birth, the sphenoidal angle opens, the occipital rises and the basi-cranio-facial balance moves away from the vertical.

Facial growth in great apes retains class I occlusion while the base grows in extension. The absence of retromandibulia is therefore explained by canine lock. The dentition participates in the balance between cranio-spine and cranio-facial fields thanks to this lock, the posture evolves from a bipedal balance, less advanced than in humans, to a quadrupedal balance. Occlusion is therefore involved in postural balance throughout post-natal development.

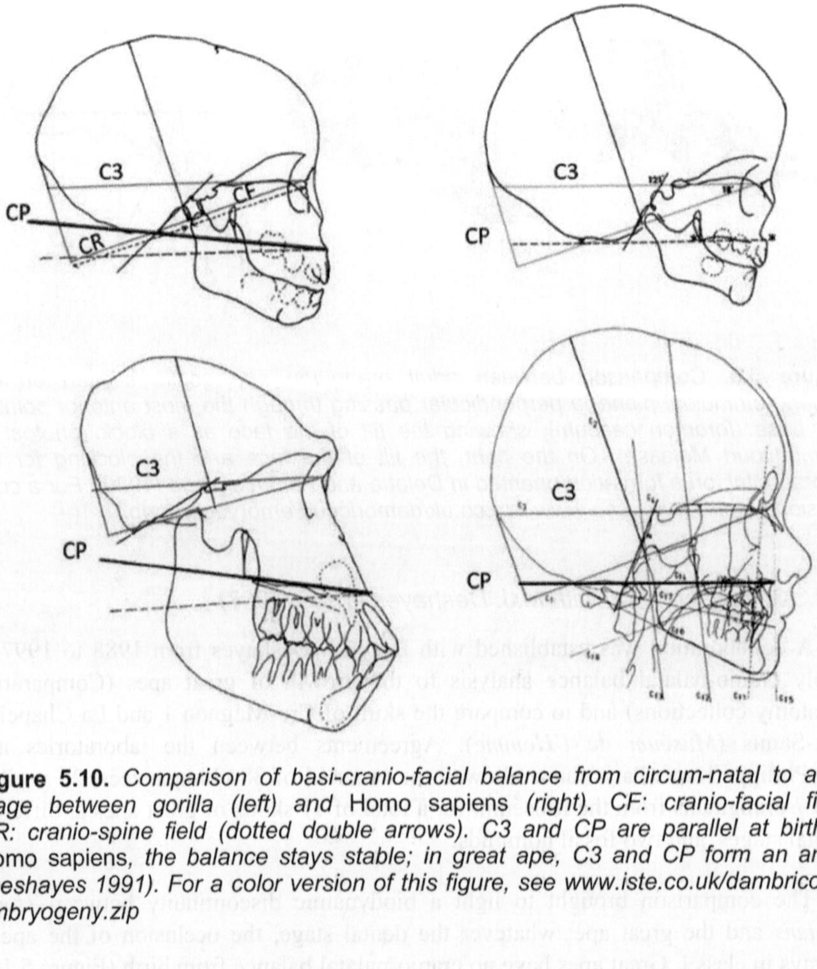

Figure 5.10. *Comparison of basi-cranio-facial balance from circum-natal to adult stage between gorilla (left) and* Homo sapiens *(right). CF: cranio-facial field, CR: cranio-spine field (dotted double arrows). C3 and CP are parallel at birth in* Homo sapiens, *the balance stays stable; in great ape, C3 and CP form an angle (Deshayes 1991). For a color version of this figure, see www.iste.co.uk/dambricourt/embryogeny.zip*

The divergence of the orangutan

Cephalometric analysis highlights the tilting movement of the face relative to the paninae (Deshayes 1991). The spheno-ethmoidal and palatal planes are parallel as in *Homo sapiens*, but this one does not extend to the cerebellar fossa. The cranio-palatal equilibrium does not exist and the spheno-occipital trajectories remain the same as that of the gorilla; the sphenoidal angle opens and the occiput moves upward (Figure 5.11).

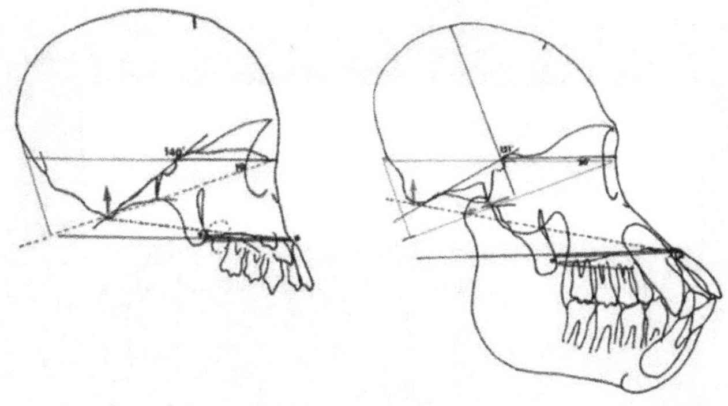

Figure 5.11. *Comparison of the basi-craniofacial equilibrium of the orangutan's temporary and adult dentition. The palatal plane (between the two asterisks) is parallel to the spheno-ethmoidal plane, and the cranio-palatal line goes up toward the occipital and moves away from the palate as it grows (Deshayes 1991). For a color version of this figure, see www.iste.co.uk/dambricourt/ embryogeny.zip*

The parallelism of the palatal and sphenoidal planes is due to the tilting movement of the face. The cranio-facial and cranio-spine fields remain equivalent. This equilibrium is enabled by the canine lock and the activity of the spheno-ethmoidal synchondrosis which is still active, whereas it is synostosed in the paninae (Deshayes 1991). The growth of the anterior base is influenced by these kinetics. Thus, ponginae and paninae have identical mandibular and spheno-occipital morphogenesis, however, the activity of the spheno-ethmoidal synchondrosis differs and the trajectories of the orangutan's face move upward.

5.3. The embryonic origin of the sphenoidal angle and cranio-spinal straightening

5.3.1. *Chondrocranium rotation according to Levi (1900)*

The only skull that forms in the embryo is the base. It is cartilaginous, hence its name, the chondrocranium. The vault and the occiput do not exist since the cerebral hemispheres and the cerebellum are not formed. At the earliest stages of embryogenesis, the future brain is reduced to the cephalic neural tube with three vesicles: the forebrain, the midbrain and the hindbrain, aligned with the future spinal cord. Beneath the neural tube extends the dorsal cord. At this stage, the human embryo follows the primitive organization of vertebrates or chordates, along an antero-posterior or cranio-caudal (also called cranio-spinal) axis (Figure 5.12).

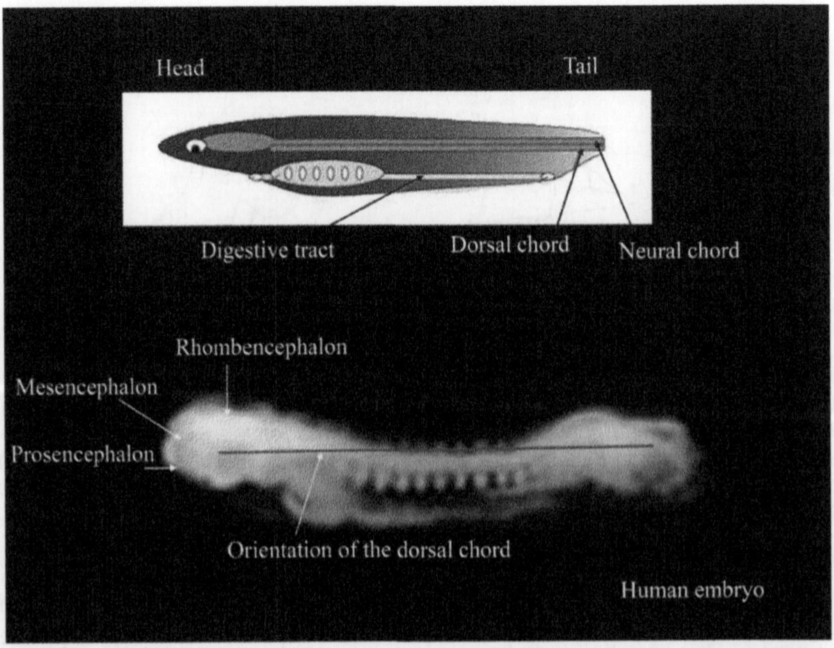

Figure 5.12. *Primitive organization of vertebrates and the 4 weeks old human embryo (3 mm, Carnegie Stage 11, MRI Kohei Shiota, Kyoto University). For a color version of this figure, see www.iste.co.uk/dambricourt/embryogeny.zip*

The chondrocranium is located under the rhombencephalon and is formed by the pre-sphenoid, the post-sphenoid and the basi-occipital. These are cartilage nuclei separated by synchondroses. They form the *planum basale* (Figure 5.13, stages 16–17). Studies on its formation are very rare, as shown by the few quotes from Müller and O'Rahilly, who have defined 23 stages of the Carnegie collection (Carnegie Institution, Washington) (O'Rahilly 1975, 1981; Müller and O'Rahilly 1980a, 1980b; O'Rahilly *et al.* 1980).

During the 1980s, embryologists referred to the work of the late 19th and early 20th centuries. From 1887 to 1978, the literature only has 25 authors, including only eight between 1951 and 1978. It was necessary to go back to 1900 to find a detailed description, published by Guiseppe Levi, which included photos of remarkable quality. *His observations reveal that the original conformation is flat and not bent as Bolk had assumed. At the very beginning, the sphenoidal angle does not exist.* This article is the only one in the entire scientific literature with the description and details of chondrocranium flexion. *It was totally unknown to paleontologists and paleoanthropologists* and it still remains so for the great majority for whom the embryogenesis of *Hominidae* does not appear as the evolutionary question.

Levi describes the steps of the plicature of the *planum basale* without giving explanations, but the photos allow us to follow the movement of the cartilaginous structures, linked to each other by connective tissue (Figure 5.13). The stages were placed in the Carnegie classification. This is defined by the crown-rump length (CRL) of the embryo, the number of days after fertilization and especially the appearance of features. Levi's stage assignment is based on his information which is limited to size, knowing that his extremes overlap two consecutive stages. The 17 mm embryo can be attributed to the end of stage 18 (13–17 mm) or to stage 19 (16–18 mm), knowing that stage 20 measures 18–22 mm. Within 1 mm, without any other data, this 17 mm embryo can be at the beginning of stage 20.

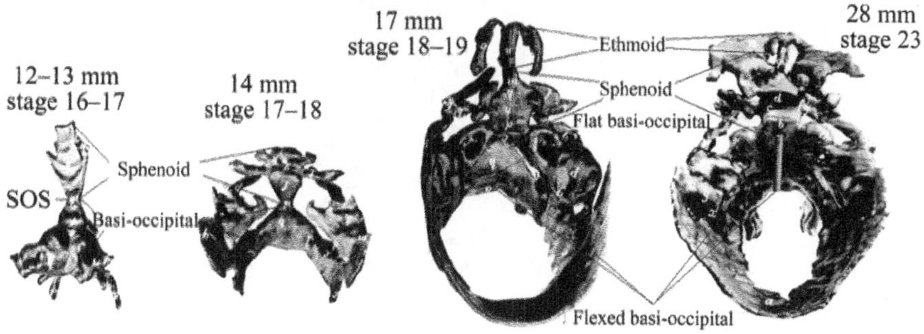

Figure 5.13. *Four stages of the human chondrocranium from above and put together for comparison, according to Levi (1900). In stages 16–17 and 17–18, the sphenoid is flat. In stages 18–19, the basi-sphenoid is verticalized, while the basi-occipital remains flat. At stage 23, the basi-occipital is aligned with the straightened basi-sphenoid. Together they form the clivus (Dambricourt Malassé 1999)*

The entirety of the structures are visible: the otic capsules containing the three semi-circular canals and the cochlea in contact with the sphenoid, the different cartilages of the occipital, and above all, Meckel's cartilage, mentioned earlier, which forms the cartilaginous pattern of the future mandible, or the first branchial arch of the splanchnocranium. The posterior end of Meckel's cartilage is attached to the almost related otic capsule of the basi-sphenoid. *In other words, in the embryo, the articulation of the mandibular cartilage is not outside the cerebral skull, but in the chondrocranium, and related to the basi-sphenoid.* Figure 5.14 shows the two cartilaginous wands seen from the bottom at the earliest stage described by Levi, pointing downward when the *planum basale* is still flat. The scale is of the order of a few millimeters.

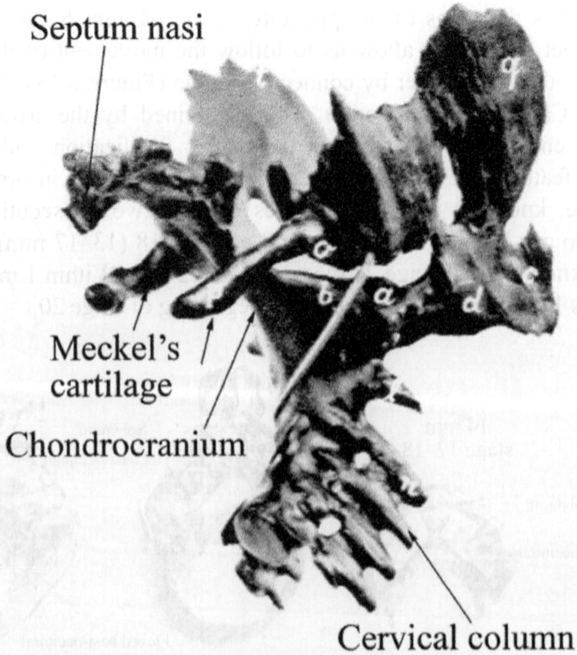

Figure 5.14. *Inferior view of a human embryo in stages 16–17. The chondrocranium is flat (Dambricourt Malassé (2005) from Levi (1900))*

In the 17–18 stages, the chondrocranium enlarges and then undergoes a two-stage deformation, as shown in the 14 mm stage in Figure 5.13. At the sixth embryonic week (stages 18–19), an antero-posterior deformation of the planum occurs without the basi-occipital and ethmoid changing their alignment. The result is a step in profile view. The ethmoid and the pre-sphenoid are raised while the basi-occipital is lowered, bringing the occipital and vertebral territories closer to the facial pole. *The chondrocranium has just acquired its verticality.* Then at stage 20, the basi-occipital is aligned with the post-sphenoid, thus forming the clivus. At stage 23, the chondrocranium has acquired its verticalized organization. Embryologists such as Kernan will confirm the conformations from stage 20 onward, having observed no earlier stages.

The 16–17 and 17–18 stages are only described by Levi. The cartilaginous areas that amortize the postero-anterior deformation of the *planum basale* are the two synchondroses that isolate the basi-sphenoid. The latter pivots on its transverse axis from top to bottom. *The sphenoidal synchondroses are the seat of the dynamical process of verticalization.*

5.3.2. *A marker for embryonic trajectories: the dorsal cord*

Levi gave no explanation for these movements. To understand them, it was necessary to put the different stages into their neural context by proceeding to a synthesis of the publications, then to define an orthonormal coordinate system. The dorsal cord has imposed itself since its cephalic extremity is embedded in the basi-sphenoid. In other words, the basi-sphenoid and the basi-occipital belong to the chordal skeleton of the embryo, while the pre-sphenoid and the ethmoid are located in front of the dorsal cord.

The horizontal axis X is given by the dorsal cord, the origin 0 is its apex and the vertical axis Y is the perpendicular passing through the origin. The protocol was published in detail in 1993 (Dambricourt Malassé 1993). The drawing in Figure 5.15 is from the American embryologist and neuroanatomist Suzanna Phelps Gage (1857–1915). The orthonormal landmark was then enough for the different stages to be published in the scientific literature and compared with the changes in position of the different neural structures and those of the cartilaginous structures described by Levi. The cell tissue above the cord, or ectoderm, first forms a neural plate. This plate elongates and its edges rise to form the neural gutter and then meet to form the neural tube (Schoewolf *et al.* 2014, p. 89). The cephalic portion is formed by three then five vesicles: the prosencephalon gives the telencephalon or cerebral hemisphere each side of the symmetric plane, and the diencephalon; and the mesencephalon does not change and the rhombencephalon differentiates into the metencephalon (future cerebellum) and myelencephalon (future spinal bulb). The cord is under the rhombencephalon.

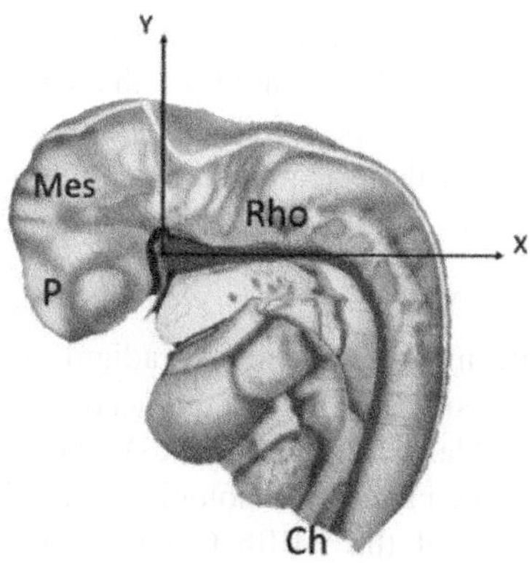

Figure 5.15. *Orthonormal landmark of the cephalic end of the neural tube at the three vesicles stage. Ch: dorsal cord, Mes: mesencephalon, P: prosencephalon, Rho: rhombencephalon. Drawing in Gage (1905) (Dambricourt Malassé 2006a). For a color version of this figure, see www.iste.co.uk/dambricourt/embryogeny.zip*

The chronology of neural tube movements and its correspondence with those of the *planum basale* are followed from stage 13 onward. At this stage, the chondrocranium is flat. *A first synthesis is illustrated in Figure 5.16. It constitutes the discovery of the process at the origin of the straightening, craniofacial contraction and verticality of hominids, still unknown in 1987.*

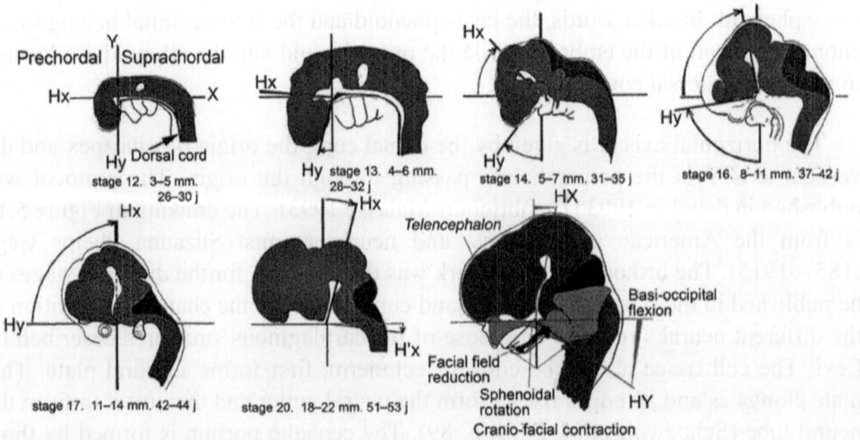

Figure 5.16. *Movement of the cephalic end of the neural tube in the orthonormal landmark. The last stage, stage 23, is from Lewis (1920) (Dambricourt Malassé 1993)*

The neural tube extends in front of the chordal apex by rotating clockwise and reaches the plumb of this termination, as shown in stage 17. At this stage, the *planum basale* is still flat. The next stage, stage 18 of the neural rotation, was not illustrated but this is the critical threshold as the basi-sphenoid straightens by pivoting on the axis, then the basi-occipital aligns, the otic capsules tilt downward and forward, and Meckel's cartilage acquires verticality in its facial extremity. This basi-cranio-facial contraction from stage 18 to stage 23 is visualized by the pantograph.

This observation, conforming to Gudin's paradigm, was immediately presented at three international symposia, the first on the head and neck in 1989, organized by the neurophysiologist Alain Berthoz (CNRS), the Rockefeller University of New York and chaired by the paleoanthropologist Phillip Tobias (Dambricourt and Deshayes 1992), the second at the XIIIth Congress of the European Society of Dentofacial Orthopaedics in 1990 (Deshayes and Dambricourt Malassé 1991) and the third in evolutionary paleontology organized by the paleontologists Jean Chaline and Lars Werdelin with the European Science Foundation in 1992 (Dambricourt Malassé 1993). The pooling of knowledge made it possible to update the interest in the application of teleradiography to the study of craniofacial hominization (Deshayes *et al.* 1990), to explore the phonatory capacities of hominids in a different way (Dambricourt Malassé and Deshayes 1995) and to present the new findings to various learned societies, both European (Dambricourt Malassé and Deshayes 1990) and Euro-American, such as the Language Origin Society. Similarly, new relationships have formed in paleontology with Jean Chaline and his team from the University of Burgundy, in order to formalize this evolutionary craniofacial process (Chaline *et al.* 1996, 1998, 2000; Dambricourt Malassé *et al.* 1999).

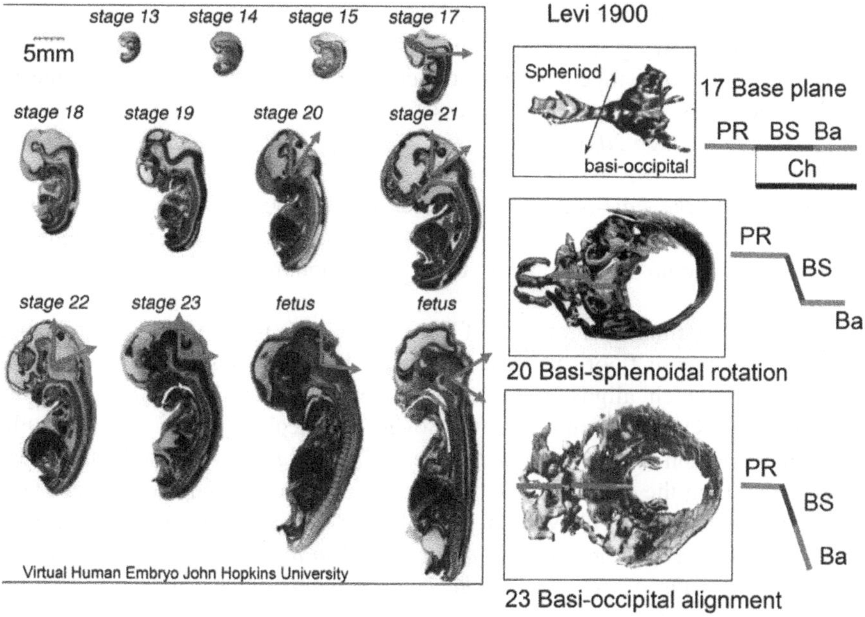

Figure 5.17. *Comparison between three stages of the human chondocranium described by Levi (1900):* planum basale *flat (stage 17), contraction-rotation of the basi-sphenoid (stage 20) and alignment of the basi-occipital (stage 21) and movements of the metencephalon and myelencephalon (Carnegie Institution micro-CT-scans of human embryos). This comparison is the first to show the relationship between the total closure of the IV ventricle (stage 20) and the underlying contraction of the* planum basale *through rotation of the basi-sphenoid (Dambricourt Malassé 2014). For a color version of this figure, see www.iste.co.uk/dambricourt/embryogeny.zip*

This synthesis of embryonic data is regularly verified as medical imaging progresses, always being adjusted and re-confirmed (Dambricourt Malassé 2011c). In 1999, researchers from Hopkins University carried out an animation within the Virtual Human Embryo program using human embryos and fetuses from the

Carnegie collection, scanned with the magnetic resonance imaging (MRI) method (Paidas *et al.* 1999; Smith 1999). The latest synthesis has been enriched and is becoming more precise. Figure 5.17 highlights more precisely the neural phenomenon that generates the rotation of the basi-sphenoid. The critical threshold is stage 20.

Older articles did not allow for details to be tracked. In fact, the floor of the rhombencephalon (or the IV ventricle, filled with cerebrospinal fluid) performs a double movement. The segment that extends in front of the cord (metencephalon) migrates upward and backward, but the supra-chordal segment (myelencephalon) also migrates following the reverse movement from stage 18 onward. The two floors eventually come together above the chordal apex at stage 20, so that the VI ventricle becomes completely compressed. *Levi's 17 mm embryo corresponds most probably to stage 20, with the basi-sphenoid pivoted on its axis, between two synchondroses, the one that separates it from the pre-sphenoid and the one that isolates it from the basi-occipital.* In other words, the transverse axis, around which the *basal planum* bends and contracts, is in the intra-sphenoidal synchondrosis and not in the spheno-occipital synchondrosis.

After this maximum contraction of the IV ventricle, the movement is reversed only for the posterior floor (myelencephalon), which then descends and drags the basi-occipital, which aligns with the basi-sphenoid thanks to spheno-occipital synchondrosis. Once the movement is over, the cephalic pole is still in the embryonic position coiled around the heart. It straightens and a human head emerges, marking the end of the embryonic period. The telencephalon is not very developed and in front of the vertical given by the Y-axis in the center of the base. *The growth of cerebral hemispheres and cerebellum therefore have no role in the straightening of the basi-sphenoid, the basi-occipital and the tilting of the otic capsules down and forward.*

The converging movements of the metencephalon and myelencephalon are due to the migration of the neural crest, the cells that form the ceiling of the neural tube. They migrate toward the ventral pole by widening the tube on the sides, leaving only a membrane to form the roof of the IV ventricle. The cells are connected by fibrils which form a matrix, the tensions exerted on either side of the chordal termination are maximal and bring the metencephalon and myelencephalon together (Figure 5.18).

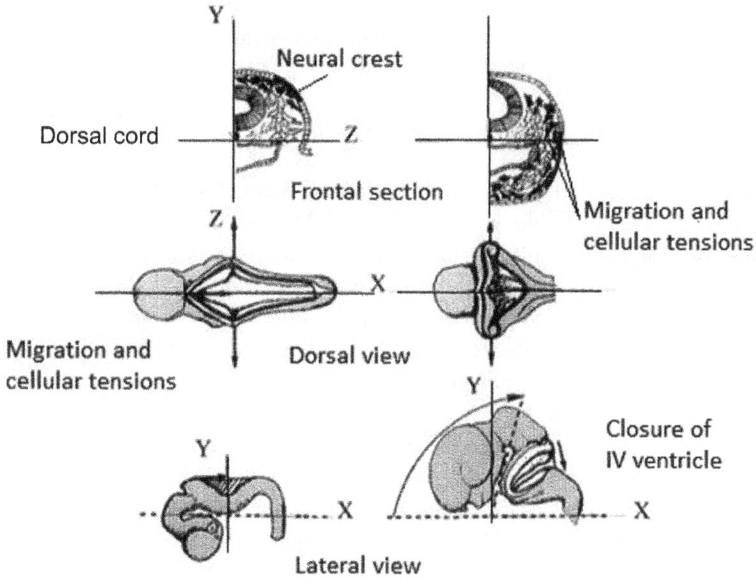

Figure 5.18. *Three-plane view of the relationship between the migration of the neural crests above the cephalic termination of the dorsal cord and the closure of the IV ventricle. The diagrams are taken from Langman (1984). Vectors and orthonormal landmarks have been added (Dambricourt Malassé 1999)*

Experiments carried out on newts in the 1970s showed that since the embryo is 95% water, the summation of neuroepithelial dynamics correspond to viscoelastic properties. The neuroepithelial cells (cells of the neural tube) have inner microfilaments and contract differently all along the neural tube; they are also connected by fibers that bind them together, and they contract to a maximum if the neuroepithelium is in contact with the dorsal cord (Jacobson and Gordon 1976; Jacobson 1978). Those viscoelasticity properties have been confirmed in mice (Chengxing *et al.* 2018) and the presence of hydrogel in the extracellular matrix has now been shown to be critical in neuron formation processes (Moxon *et al.* 2019; George *et al.* 2020). The animation by the Virtual Human Embryo program shows the movements of the cephalic neural tube. There seems to be an alternation of closing and releasing phases like pulses, but stage 20 is the maximum closure of the IV ventricle and the rotation of the basi-sphenoid occurs at this physical limit. Finally, a comparison of the animation with the embryonic stages of mice shows that stage 20 and subsequent stages are absent without the straightening of the brain stem (Wong *et al.* 2015). The movements of the IV ventricle of mice are part of a completely different cranio-spinal dynamic which ceases earlier. *Human rotation is thus the cephalic effect of tensions during the elongation of the embryo that can hardly be explained without fiber networks which extend into the synchondroses.*

5.3.3. *Morphogenesis of Meckel's cartilage*

Louis Bolk described the different stages of Meckel's cartilage formation and ossification of the mandible well. Lewis and Macklin's plates completed Levi's descriptions. The combination of all these illustrations makes it possible to follow the correlations of displacements as soon as the *planum basale* begin to fold. The rotation of the basi-sphenoid pulls down and forward the two large otic capsules and thus the posterior end of Meckel's cartilage. The two wands grow in length and progressively join in the sagittal plane to form an arch, in a context constrained by the sphenoidal rotation (Figure 5.19). Gradually, the arch straightens and forms a hook, the *hamulus meckeli*. When the rotation has reached its end, the ventral end of the mandibular cartilage is straightened with an underlying triangular gap in frontal view. The membranous ossification follows the cartilaginous skeleton and its straightened termination gives the symphyseal vacuum absent in monkeys and great apes.

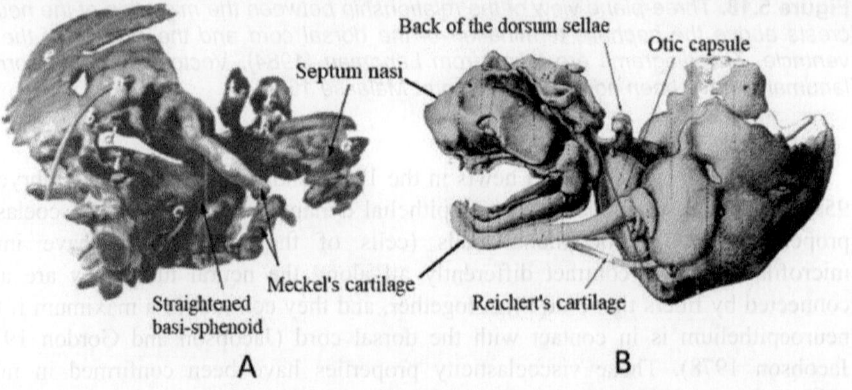

Figure 5.19. *Levi embryo at stages 18–19 and 40 mm fetus by Macklin (1914) showing the ventral straightening of the mandibular arch formed by the union of the two Meckel cartilages and their dorsal connection with the straightened otic capsules (Dambricourt Malassé 2005, 2006)*

Thus, the structure of the mandibular symphysis and the symphysis angle, which distinguish hominins from great apes, are set up at the end of the embryonic period, in relation to the amplitude of basi-sphenoidal rotation, which is a consequence of ventricle IV's closure.

5.3.4. *The organization of intra-sphenoidal synchondrosis: a convection cell*

Symmetrical coordination is observed between embryonic growth vectors of the face and of the skull base relative to a frontal plane passing through the intra-sphenoidal synchondrosis. These processes would be impossible without the existence of a tissue network with visco-elastic mechanical properties in a gelatinous matrix. Developmental genetics would be ineffective without water, both inside and outside the sex cells; three days after fertilization, the embryo is 97% water. We find here the concept of self-organized dynamic systems. By 2008, modeling had pushed investigations further (Xiaoguang *et al.* 2008) and dynamic simulations were available on the sites of research teams (Broadland *et al.* 2010), in particular the studies of Vincent Fleury (Fleury 2011). This network of filamentous molecules of the extra-cellular matrix covering the entire organism from the molecular level to the tissues has gradually become established (Guimberteau and Armstrong 2016). It is thus during embryogenesis that the shapes of the skeleton and topographic relationships between its different parts could change following the cephalocaudal gradient. The head is differentiated before the post-cranial skeleton. The phalanges of the hands are already distinct while those of the feet are still paddles. The shoulder blades are in a lateral position as in a quadruped, but when the head is straightened, releasing the neck, the tensions modify the position of the shoulders and the forelimbs are in a different topographical relationship from that of a great ape, whose neural kinematics do not have the same amplitudes and tensions.

To account for the embryonic rotation axes, the synchondroses of the sphenoid must behave like fluids.

In 1989, the physiology of intra-sphenoidal synchondrosis was not known and the possibility that it could be the seat of constraints generating rotations was not on the agenda. The Fyssen Foundation[3] supported the post-doctoral research. The forensic medicine laboratory of the Raymond-Poincaré de Garches Hospital (part of the Hospital Trust of Greater Paris) entrusted the inventory of 65 human fetuses whose sex and age had to be identified by the crown-rump length (Dambricourt Malassé 1989). A 4-month-old fetus was mounted on a slide in order to observe the intra-sphenoidal synchondrosis. This is composed of collagen fibers whose spatial organization corresponds to two rotations in opposite directions, two "convection currents" typical of fluids. The cells of cartilaginous tissue (chondroblasts) are located on each side of the physical limit between the two territories (Figure 5.20). *This spatial organization corresponds exactly to the prediction of dynamics in*

3 The French FYSSEN Foundation aims to promote the scientific analysis of the logical mechanisms of animal and human behavior and their ontogenetic and phylogenetic development in all its forms.

rotation and symmetrically with respect to a transverse axis passing through this synchondrosis, generating the cranio-facial contraction, between the cranio-facial field and the cranio-spinal field.

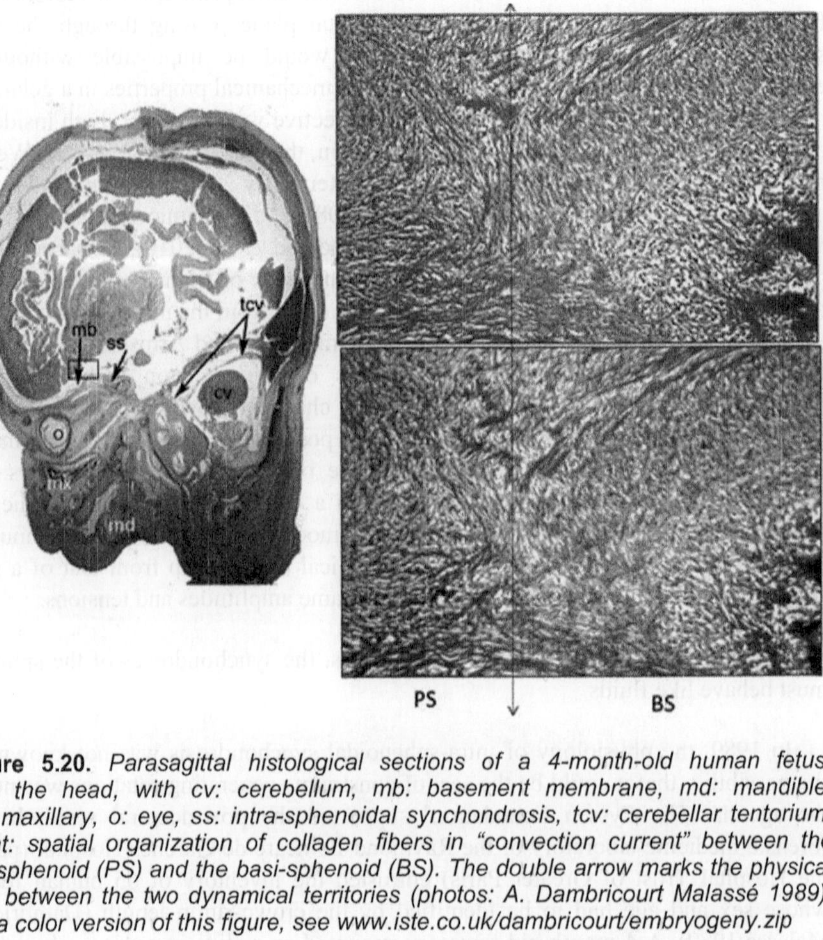

Figure 5.20. *Parasagittal histological sections of a 4-month-old human fetus. Left: the head, with cv: cerebellum, mb: basement membrane, md: mandible, mx: maxillary, o: eye, ss: intra-sphenoidal synchondrosis, tcv: cerebellar tentorium. Right: spatial organization of collagen fibers in "convection current" between the pre-sphenoid (PS) and the basi-sphenoid (BS). The double arrow marks the physical limit between the two dynamical territories (photos: A. Dambricourt Malassé 1989). For a color version of this figure, see www.iste.co.uk/dambricourt/embryogeny.zip*

The separation between the two territories of the skull base, cranio-facial and cranio-spinal, is therefore formed by millions of collagen fibers organized in the "convection current". Chondroblast genes allow the synthesis of collagen fibers but they do not code for their orientation. The observation noted a typical structure of a hydrodynamic system. This histological section made it possible to infer kinetic and dynamic fields coordinating the balances between the base tensions and the growth vectors of the face. *The growth vectors of the synchondroses are rotations and, thanks to these rotations, the growing cranial puzzle develops in a coordinated way between the flexion of the base and the contracting face. Without this organization of the collagen fibers, the flexural morphogenesis of the skull would be impossible, because the mandible propelled downward and forward would no longer be able to coordinate with the overlying maxillary bud.* The "all genetic" paradigm was thus invalidated, and the morphogenesis is also influenced by biophysical constraints specific to different media, viscous and solid; variables according to the scales considered. In 2004, Scott Gilbert described the return to the knowledge of the early 20th century, rich in comparative anatomy and concepts, but eclipsed by the overly exclusive doctrine of population genetics:

> The rediscovery of the morphogenetic field: one of the most fertile concepts in embryology at the beginning of the 20th century, the morphogenetic fields were somewhat overshadowed by the discovery of the role of genes in development, but we are in the process of "rediscovering" them as a result of these same genetic studies. (Gilbert, *Biologie du développement*, 2004, author's translation)

5.3.5. *The* planum basale *of other primates and mammals*

The formation of the neural tube and its pre-chordal elongation is common to all mammals; the distinctions are based on kinetics, or the duration of the fibrillar protein tensions that cause the movements of the neural tube. Mammals other than simians have not exceeded the threshold of fourth ventricular contraction, which is amplified in simians and provokes the basisphenoidal rotation.

Primate embryos and fetuses are rare. One of the best supplied collections is kept at the Hubrecht Laboratorium in Utrecht in the Netherlands. *Tarsius* embryos were observed in 1989 thanks to the Fyssen Foundation's post-doctoral fellowship, but they were not mounted on slides and MRI did not yet exist at that point. It was therefore necessary to make do with the current state of the national and international collections and to gather imaging banks from the scientific literature. Statistics do not exist for the comparative embryology of rare species, taxa are described according to the number of specimens preserved, and the synthesis of the growth trajectories of a taxon, in this case that of the mandible, enables one to conclude the embryonic grade of craniofacial contraction. It is therefore divided into four groups of increasing amplitude in present-day primate species since rank zero: (1) the absence of rotation in prosimians; (2) monkeys; (3) great apes (paninae, ponginae); and (4) *Homo sapiens* (Figure 5.21).

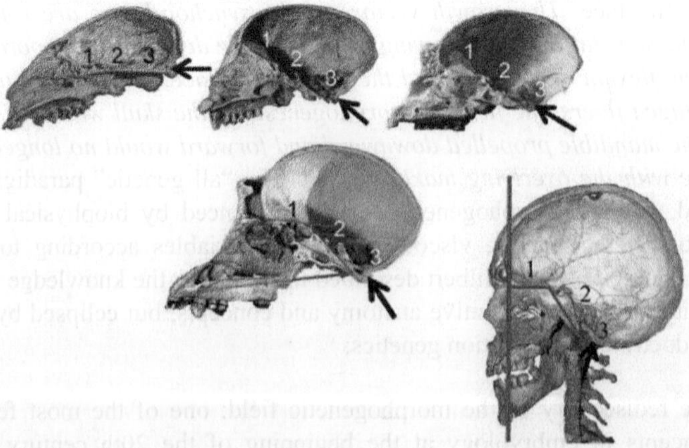

Figure 5.21. *Median sagittal section of adult skulls of current species showing the staggering of the three brain fossae: top left to right: prosimian (Indri), platyrrhines simian (Cebus), catarhines simian (Hylobates). Bottom: great ape (Pan) and* Homo sapiens. *The arrow indicates the orientation of the occipital hole (photos: A. Dambricourt Malassé). For a color version of this figure, see www.iste.co.uk/dambricourt/embryogeny.zip*

5.3.5.1. *The sloth, an arboreal mammal of tropical America*

Figure 5.22 compares the fetus of a sloth (*Bradypus*) and *Homo sapiens* showing the absence of basi-sphenoid pivoting, whereas in humans the contraction of the spheno-occipital and spheno-facial fields has reduced maxillary prognathism.

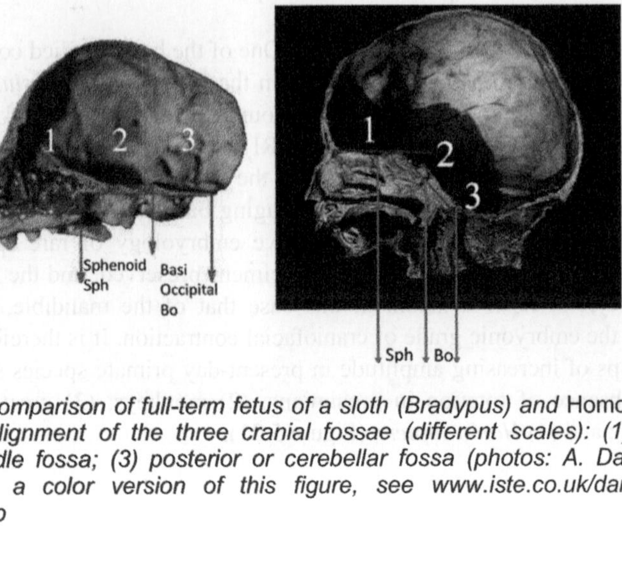

Figure 5.22. *Comparison of full-term fetus of a sloth (Bradypus) and* Homo sapiens, *showing the alignment of the three cranial fossae (different scales): (1) anterior fossa; (2) middle fossa; (3) posterior or cerebellar fossa (photos: A. Dambricourt Malassé). For a color version of this figure, see www.iste.co.uk/dambricourt/embryogeny.zip*

5.3.5.2. *The prosimians*

The mandibles of tarsiers and a Strepsirrhini like the *loris* do not differ; both are long, narrow and with a V-shape (Figure 5.23). Strepsirhini and tarsiers have no basi-sphenoidal rotation (Figure 5.24). *The chondrocranium of the tarsier is as flat as that of a non-simian mammal.*

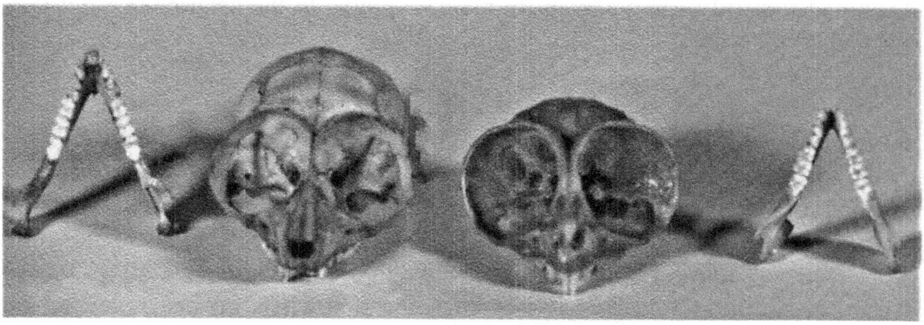

Figure 5.23. *Mandibles and adult skulls of loris and tarsier (photo: A. Dambricourt Malassé)*

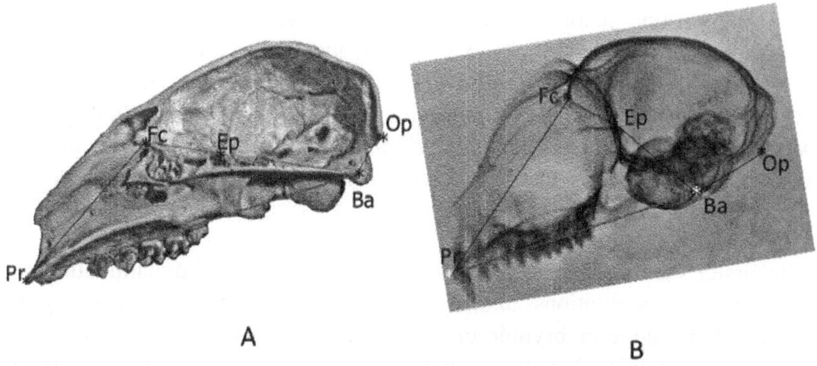

Figure 5.24. *Adult skulls of A: Indri (lemur) in sagittal section, and B: Tarsius (X-rays). Comparison of the orientation of the occipital hole, the base and the basion-prosthion straight line (red line). Ba: basion, Ep: ephippion, Fc: foramen caecum, Op: opisthion, Pr: prosthion. The skulls are oriented according to Ba–Pr parallel lines. The occipital hole (Ba–Op) of the Tarsier is not directed backward as in a lemur but it does not correspond to a lowering of the cerebellar fossa due to the pivoting of the basi-sphenoid like a simian (photo: A. Dambricourt Malassé and X-ray by Gudin 1951). For a color version of this figure, see www.iste.co.uk/dambricourt/embryogeny.zip*

From an external view, the skull gives the impression of less prognathism, but X-rays show that it does not differ from a Strepsirrhini like the *indri*, the upper face projection is masked by the hypertrophied orbital cavities. These give the impression of being placed in the frontal plane like that of simians, but this is only an effect of their lateral dilatation, whereas the simian orbital frontalization is due to the lateral expansion of the anterior cranial fossa. The cerebellar fossa also seems to be tilted down and forward as in a simian. In reality, relative to the anterior fossa, it is not lowered.

The skull of the tarsier is that of a prosimian embryo deformed by the development of the optic vesicles flanking each side of the diencephalon, as shown by the external observations in 1989. Located in front of the cord, the mesenchymal tensions act within the synchondrosis between the pre-sphenoid and the ethmoid and have nothing to do with the closure of the fourth ventricle. The kinematics of the cephalic neural tube did not exceed a dynamic threshold common to all mammals – except the Simiiforme. Twenty years later, the first MRIs of tarsier fetuses were performed by Jeffery and colleagues on a series of 19 specimens from the Hubrecht Laboratorium. The sections confirm that the chondrocranium is flat (Figure 3a in Jeffery *et al.* 2007). The adult exocranial appearances evoke similarities with simians, but they are not homologies and the embryology confirms this.

The Haplorrhine taxonomic category which separates the tarsiers from the Strepsirrhini, and brings them closer to the simians, retains only adult characteristics such as the absence of a nasal truffle. But these criteria are isolated from embryogenesis, which is the first in phylogenesis, so this classification ignores the convergences. *Embryology ultimately demonstrates the arbitrary nature of the classification which does not take into account the chronology of development starting with embryogenesis.*

The tarsier's embryogenesis does not go beyond the neural dynamic limits of the ancestral primate populations, and like the Strepsirrhini, both lineages stay within the limits of the same embryonic organizational plan, whereas the simians are no longer species of this plan, they emerged from a distinct prosimian lineage, as Teilhard de Chardin concluded. A third "pre" simian lineage is conceivable in the Eocene. The Haplorrhine category therefore has no biological reality; it was created in a gradualist and linear spirit, which does not conceive of emergence, threshold effects or macroscopic morphological discontinuities because of the still latent metaphysical conflicts of the 19th century.

5.3.5.3. *Monkeys and great apes*

Catarhini: *Cercopithecidae*

The embryonic development of the macaque has been described externally in a broad outline (Gribnau and Geijsberts 1981), allowing for comparison of the neural tube trajectories, but the descriptions are not sufficiently precise for the closure of the IV ventricle, since the *Cercopithecidae* shows a sphenoidal rotation much less than *Homo sapiens* and chimpanzees. The difference is confirmed by a very young fetus of 53 mm crown-rump length: the *hamulus meckeli* does not exist and the posterior end of Meckel's cartilage is well above the tangent to the palate (Figure 5.25).

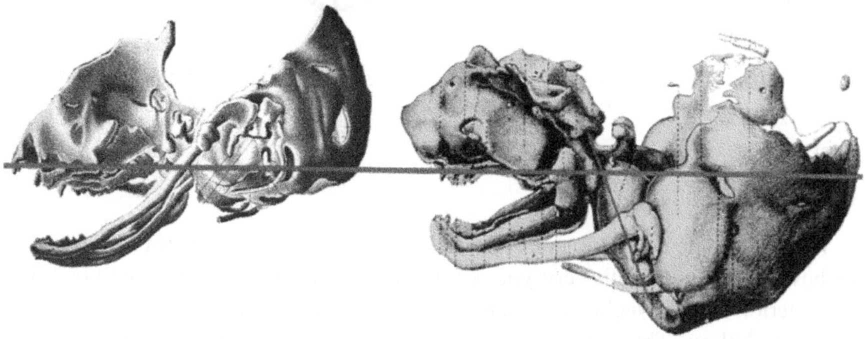

Figure 5.25. *Chondrocranium with the mandible of a fetus of* Cercopithecidae (Semnopithecus maurus)*, 53 mm CRL according to Fischer (1902) and 40 mm* Homo sapiens *according to Macklin (1914). The horizontal line is the tangent to the bottom of the septum nasi (Dambricourt Malassé 1987, 2006a). For a color version of this figure, see www.iste.co.uk/dambricourt/embryogeny.zip*

Platyrrhini

The Central American *Alouatta*, or howler monkey, is a special case that demonstrates the biodynamic unity of the base and the face. The mandible shares the common features with Catarhinians; an upsiloid alveolar arch and a post-natal basilar angular deviation lower than 60°, but the adult stage has no flexion, the cerebellum aligns with the frontal lobes and the edge of the petrous pyramids is above the spheno-ethmoidal plane. This has never been a counterexample of phylogenetic cranio-facial contraction. As in all monkeys, the sphenoid angle opens after birth, but with *Alouatta* the base has to contend with the formation of a large laryngeal sac beneath the mandible, which alters the growth paths of the upper surface. Comparison with the juvenile stage shows a rotation of the maxillary in the opposite direction of the cerebellar fossa: the posterior end of the palate tilts upward, while the anterior end migrates upward (Figure 5.25). The entire basi-cranio-maxillo-facial structure adapts to the growth of the laryngeal sac; the deformity has already begun in the young.

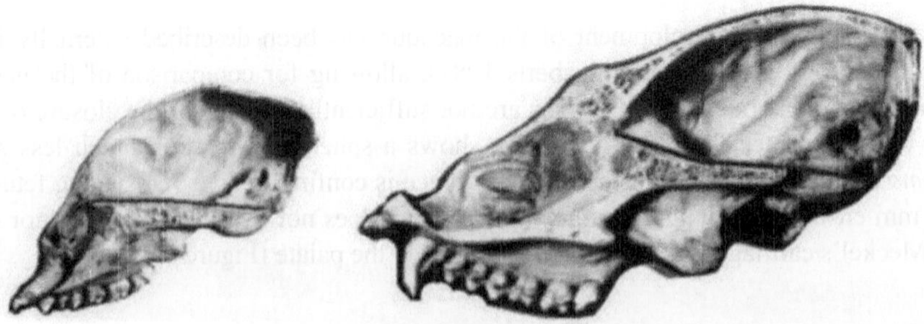

Figure 5.26. *Alouatta: comparison of the skull base angulation between the juvenile and adult stages (photo: A. Dambricourt Malassé)*

Great ape: the chimpanzee

The embryonic development of *paninea* and *ponginea* has never been described. We have to make do with very young fetuses like Starck's 71 mm. The position of the posterior end of Meckel's cartilage cuts the tangent to the palate, whereas *Homo sapiens* is below. The lowering of the otic capsule relative to monkeys such as *Cercopithecidae* is undeniable, but it is still largely insufficient compared to *Homo sapiens*, which has a completely lowered cerebellar fossa (Figure 5.27).

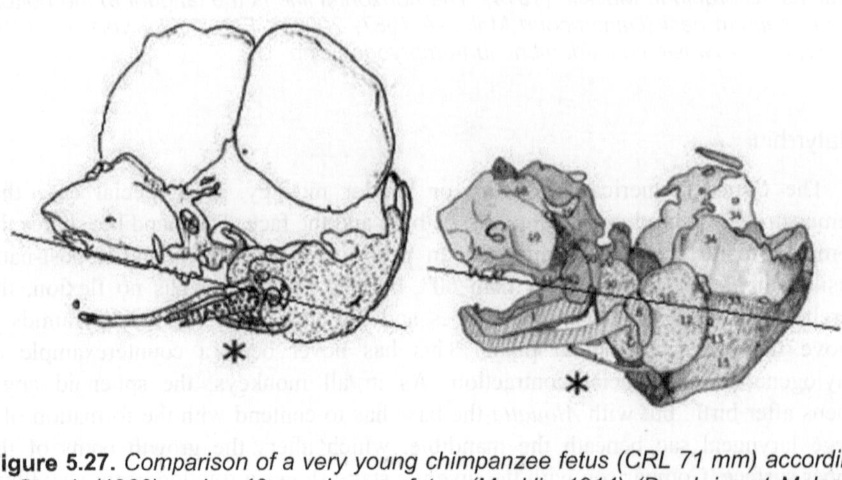

Figure 5.27. *Comparison of a very young chimpanzee fetus (CRL 71 mm) according to Starck (1960) and a 40 mm human fetus (Macklin 1914) (Dambricourt Malassé 1987). The asterisk under Meckel's cartilage highlights the difference in the amplitude of lowering of the human cerebellar fossa at the end of embryogenesis*

Embryonic data are limited in quantity, but the lowering of the prosimian cerebellar fossa can only be explained by the process we have described with *Homo sapiens* and the Sapiens' embryonic contraction can only be explained by an evolution of this embryonic process since prosimians. Thus, the differences in craniofacial contraction between prosimians, monkeys, great apes and *Homo sapiens* are genuinely embryonic in origin, the acquisition of the vertical dimension of the craniofacial structures and nervous system is a difference in the dynamics of the neural tube and that of the fourth ventricle in particular which amplify between two dynamical organizations. We have no doubt that the advancement of non-invasive medical imaging and future comparisons of primate embryos will confirm these findings.

5.3.6. *Sphenoid rotation and semi-circular channels*

New technologies in medical imaging have made it possible to explore the inner ear, located in the petrous pyramids (Figure 5.28) and formed by the labyrinth, composed of the cochlea and the semi-circular canals, which are essential for the cerebellum to control balance and head movements.

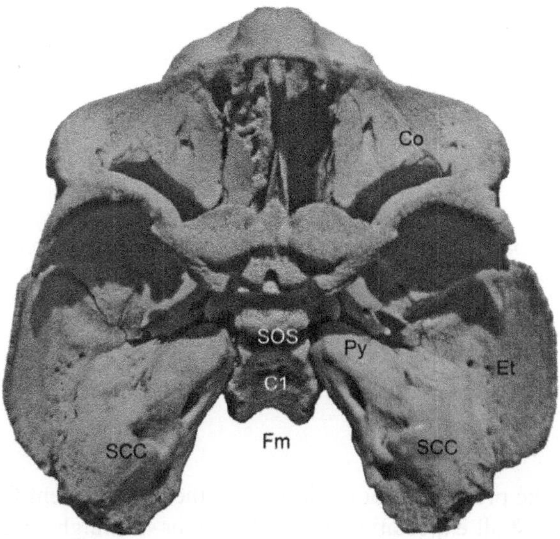

Figure 5.28. *Internal view of the cranial base of a newborn* Homo sapiens. *The superior semi-circular canal (SCC) appears in outcrop. The anterior end of the pyramids (Py) is in contact with the spheno-occipital synchondrosis (SOS). Cl: clivus, Co: orbital cavity, Et: temporal squamous, Fm: foramen magnum (occipital hole) (photo: A. Dambricourt Malassé). For a color version of this figure, see www.iste.co.uk/dambricourt/embryogeny.zip*

Delattre and Fenart (1960) found a relationship between the forward and downward rotation of the petrous pyramids and the positions of the long axis of the cochlea and the posterior semi-circular canal. They follow the clockwise direction (Figure 5.29(A) and 5.29(B)). Some 30 years later, Wind and Zonneveld (1989), Spoor *et al.* (1994) and Spoor and Zonneveld (1998) applied MRIs to great apes and fossil hominids, adding the transverse dimension (Figure 5.29(D)). They applied a morphological protocol and found a clockwise tilt of the cochlea and posterior canal between chimpanzees and *Homo sapiens*. But they also observed an angular deviation between the great axes of the lateral (horizontal) and posterior semi-circular canals, which is clearly more open in humans than in apes (Figure 5.29(D)). This opening evokes the frontalization of the two petrous pyramids. The authors then looked for correlations with the straightening of the clivus. In other words, a fossil with an upright clivus must have petrous pyramids that are also tilted downward and forward and a deviation between these two semi-circular canals as in *Homo sapiens*. These transformations of the skull are seen as an adaptation to permanent bipedalism, always seen as the locomotor engine of the hominization process.

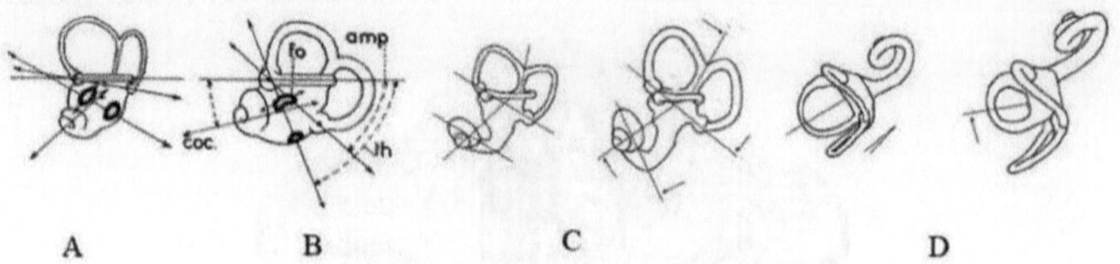

Figure 5.29. *Comparison of the position of the posterior semi-circular canal and the axis of the cochlea (coc.) oriented in the plane of the lateral semi-circular canal. A and B in lateral view: the fox and* Homo sapiens *according to Delattre and Fenart (1960); C: the chimpanzee (left) and* Homo sapiens *(right) in lateral view; and D: the same in upper view. In* Homo sapiens, *the cochlea is tilted downward and forward, and the angle between the great axes of the lateral and posterior semi-circular canals is open (Spoor and Zonneveld 1998)*

In applying the protocol to hominid fossils, the authors went from one surprise to another. Against all expectation, while the clivus is straightened, they noted the lack of angular opening between the lateral and posterior semi-circular canals in *Australopithecus* (Spoor and Zonneveld 1997), whereas it is present on *Homo* fossil species (Spoor and Zonneveld 1998). Then, in the Neanderthal, the posterior canal is higher than in the older *Homo erectus*. The simian conformation of

Australopithecus was unexpected, as it seemed logical to associate the greater angular deviation with the clivus straightening thanks to micromechanical effects and the bipedal locomotion as the major evolutionary engine. Similarly, the highest position of the Neanderthal's posterior canal does not fit with the new dogma according to which its bipedalism cannot differ from that of *Homo sapiens*; to do otherwise would be to adopt a retrograde position that would bring it closer to the ape.

In a preliminary study of fossil hominid labyrinths, we have demonstrated that *Homo erectus* shows modern human canal dimensions. In contrast, South African australopithecines show the ancestral great ape-like condition, which has been interpreted as support for studies, concluding that these hominids were facultative rather than obligatory bipeds. Subsequently, it was discovered that Neanderthals, as habitual bipeds, nevertheless have smaller vertical canals than modern humans (Hublin *et al.* 1996). Hence, the evolutionary history of the hominid semi-circular canal system is clearly more complex than a single change from a more great ape-like to a modern human-like morphology, and any link with locomotion exceeds a simple association with the broad categories of quadrupedal vs. bipedal behavior (Spoor and Zonneveld 1998).

Indeed, paleoanthropology, dominated by the dogma of locomotion, has not yet perceived that the problem is the organogenesis and embryogenesis of the nervous system. The Neanderthal was bipedal, but its occluso-postural balance was not that of *Homo sapiens*, because the embryonic straightening was less advanced.

In 1999, MRI became more accessible; it was therefore possible to verify whether these differences between the morphogenesis of the inner ear were caused by the rotation of the basi-sphenoid and the frontalization of the petrous pyramids. A biometric protocol was therefore developed in collaboration with Dr. Eric de Kerviler and Dr. Jean-Pascal Martin of the Saint-Louis Hospital in Paris (Dambricourt Malassé *et al.* 2000a, 2000b). Three angles were defined to test the correlations between the orientation of the three semi-circular canals, the sphenoidal angle and the frontalization of the petrous pyramids. These angles are the orientation of the semi-circular canals to the sagittal plane (A1 and A3) and the angular deviation between the lateral and posterior semi-circular canals (A2) (Figures 5.30 and 5.31).

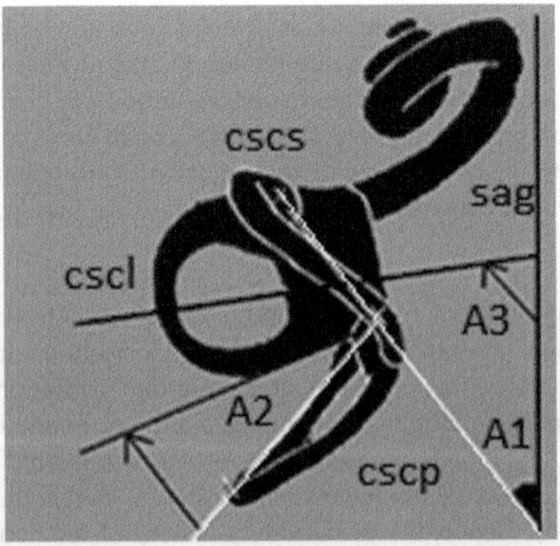

Figure 5.30. *Trace in upper view of the inner ear established on the graph by Spoor and Zonneveld (1998). Angles have been added: A1: between the sagittal plane and the major axis of the superior semi-circular canal; A2: between the major axis of the posterior semi-circular canal and the posterior tangent of the lateral semi-circular canal; A3: between the sagittal plane and the major axis of the lateral semi-circular canal; cscl: lateral semi-circular canal, cscp : posterior semi-circular canal, cscs : superior semi-circular canal, sag : sagittal plane (Dambricourt Malassé* et al. *2000a, 2000b)*

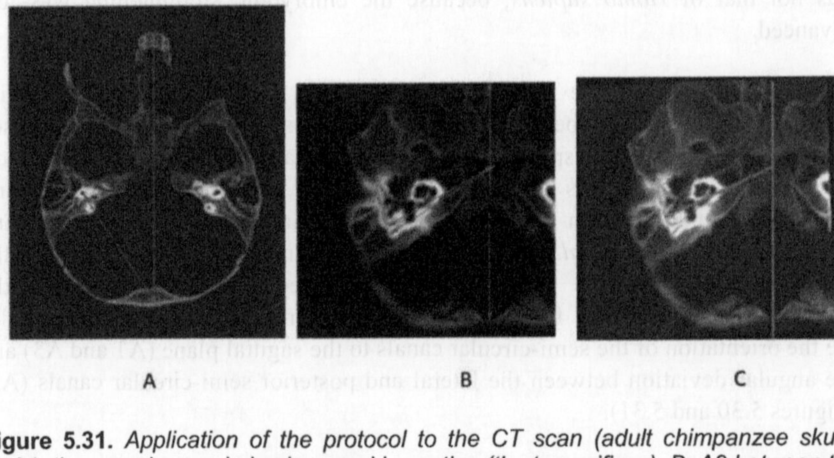

Figure 5.31. *Application of the protocol to the CT scan (adult chimpanzee skull). A: A1: the superior semi-circular canal in section (the two orifices). B: A2 between the posterior edge of the lateral semi-circular canal (shape in semi-circle) and the posterior semi-circular canal (in a fairly long section). C: A3 between the sagittal axis and the major axis of the lateral semi-circular canal (Dambricourt Malassé* et al. *2000a, 2000b). For a color version of this figure, see www.iste.co.uk/dambricourt/embryogeny.zip*

Scans were still rare in paleoanthropology. Two *Homo erectus* from Java (Ngandong 1 and Sambungmacan 1) were entrusted to Dr. Kerviler by the paleoanthropologist Teuku Jacob (1929–2007), completed by two of the oldest known *Homo sapiens* fossils, Qafzeh 6 and 7 (Israel, 95,000 years old) preserved at the IPH. Only Qafzeh 6 and Sambungmacan 1 were exploitable, Qafzeh 7 being too mineralized and Nangdong 1 insufficiently preserved. The measurements were compared to an ontogenetic series of six chimpanzees, as well as one fetus and one adult gorilla. *The tests were negative. Indeed, the angle between the two canals stay stable while the sphenoidal angle opens and that of the pyramid frontalization closes* (Dambricourt *et al.* 1999, 2000).

Still searching for an explanation of the *Australopithecus* simian conformation, Jeffery and Spoor (2004) turned their attention to brain growth, particularly that of the occipital lobe, whose vision-related functions are also important in controlling balance:

> It is the possible links with locomotor behavior, body posture, basicranial remodeling and brain development that make the comparative analysis of the human bony labyrinth of particular interest, because all of these aspects play a crucial role in human evolutionary history. (Spoor and Zonneveld 1998)

The authors then sought to test Bolk's theory of fetalization, which assumes pressure from the cerebral growth on the petrous bones, which would have formed in a straightened position. They measured the sphenoidal angle and the one between the two lateral and posterior canals, on a large series of human fetuses studied by MRI from the ninth week, that is at the end of embryogenesis. Unsurprisingly, the results invalidated the dogma of encephalic pressure as the mechanical cause of sphenoidal angulation since the cause is embryonic. *More importantly, the human deviation between the two canals is already visible at 9 weeks, at the end of embryogenesis.*

Since the rotational dynamics of the basi-sphenoid are induced by the closing tensions of the IV ventricle, it was necessary to verify whether these tensions influence the morphogenesis of the semi-circular canals. The solution to this question was to scan a human fetus without rotation; an anencephalic fetus. We expected undifferentiated canals.

5.3.7. *The base of a human fetus without straightening: the semi-circular canals are still human*

The *Musée de l'Homme* holds a collection of anencephalic human fetuses, including a unique case without basi-sphenoid flexion, aged 5 months. If the deviation between the lateral and posterior canals was a consequence of the neural tensions and rotation, it must have been absent on this fetus. With Dr. Fabienne Lallouet, we compared it to a normal *Homo sapiens* fetus from the same collection and to a gorilla fetus from the IPH, aged 5 months. We applied a metric protocol.

The sphenoidal angle does not exist and the ethmoido-maxillary block is collapsed in front of the body of the sphenoid (pre- and post-sphenoid), remaining aligned with the basi-occipital (Figures 5.32 and 5.33). The otic capsules have not tilted forward and downward. *But the human deviation between the two canals is present* (Figure 5.34(C)).

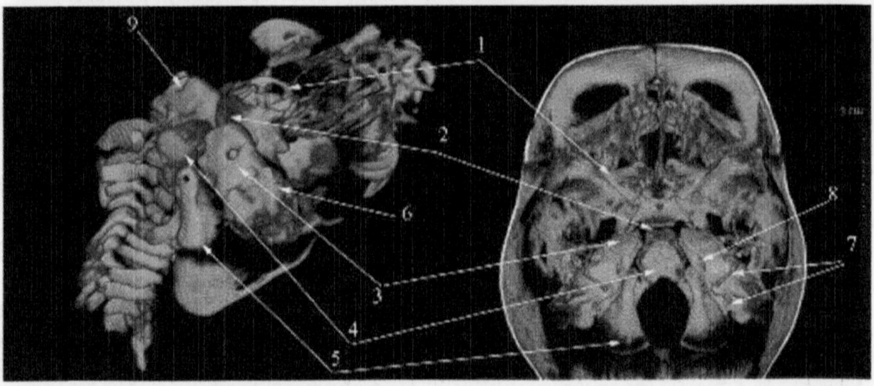

Figure 5.32. *Five-month-old human fetus. Identification of homologous characteristics of the anencephalic fetus (left) with the reference (right). (1) lesser wings of the sphenoid; (2) SOS space; (3) petrous bone; (4) basi-occipital; (5) exo-occipital; (6) location of semi-circular canals; (7) lateral and posterior canals; (8) carotid canal opening tilted toward the cerebellar fossa; (9) carotid canal opening in cranial orientation. The anencephalic fetus shows no rotation of the basi-sphenoid, no verticalization of the basi-occipital and no downward rotation of the petrous bones (photos: Dambricourt Malassé and Lallouet). For a color version of this figure, see www.iste.co.uk/dambricourt/embryogeny.zip*

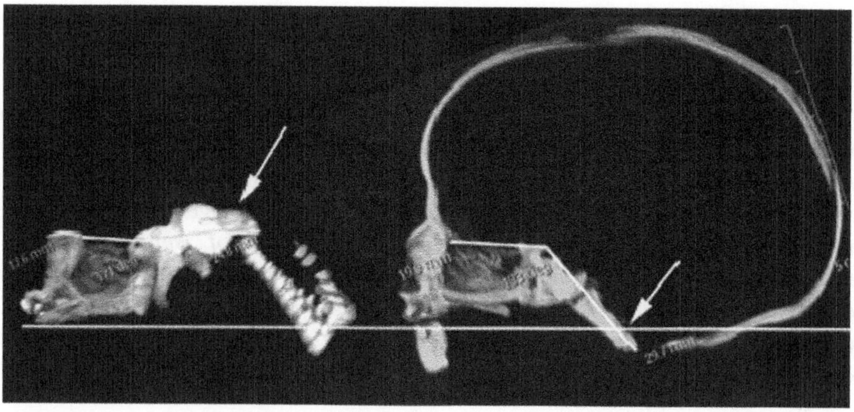

Figure 5.33. *Five-month-old human fetus. Measurement of the classical sphenoidal angle on the anencephalic (left) and reference fetus (right). The arrow indicates the basi-occipital (photos: Dambricourt Malassé and Lallouet)*

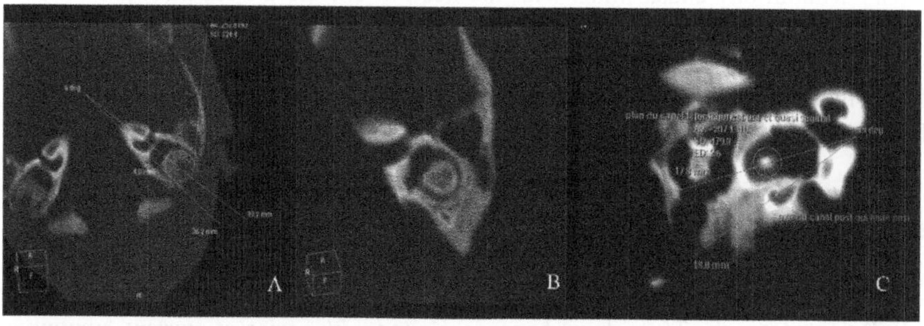

Figure 5.34. *Five-month-old gorilla and* Homo sapiens *fetus. A: the gorilla fetus, B: the* Homo sapiens *reference and C: the human anencephalic fetus. Measurement of the angular deviation between the major axis of the lateral canal and the major axis of the posterior canal. The reference plane is given by the lateral semi-circular canal's plane: the widest human angle is vizualised by the opening of the two convergent lines (C) (Dambricourt and Lallouet 2009)*

The largest angular deviation is therefore not dependent on the rotation of the basi-sphenoid and the kinetics of the neural tube. This finding confirmed the conclusions of an evolution of embryogenesis not related to locomotion but to the regulation of the cephalocaudal gradient of embryogenesis (Dambricourt Malassé 1988).

The activity of architect, or homeotic, genes was discovered later (Gehring 1999) and awarded the Nobel Prize for Physiology or Medicine in 1995, confirming the existence of molecular mechanisms capable of modifying embryogenesis along the cranio-sacral axis.

The observation of the normal formation of the two human semi-circular canals without any relation to the tensions of the neural dynamics masterfully confirmed that the origin of the Australopithecus and the first Homo species *had strictly no relation to locomotor constraints, but that the two embryogeneses resulted from very close complexification in the genetic regulation of embryogenesis. The tensions of the neural tube at the time of the closure of the IV ventricle are only one of the many consequences, or chain reactions, of a change present from fertilization causing an accentuation of the rotation of the basi-sphenoid and straightening of the embryonic axis.*

The emergence of hominids is thus indeed the consequence of an evolution of the genetics of embryonic development; *the axial straightening is only an epiphenomenon of a prolongation of neurogenesis with its complexification.* The emergence of an Australopithecine was thus to be based on the same mechanisms as *Homo* without any causal links with the need to change the locomotion. These are two axial straightenings according to different embryonic modalities with spectacular effects: a vertical nervous systems moving exclusively in support of the lower limbs and which could live in the tropical and moist forest in which they were born.

5.4. Craniofacial contraction

These conclusions about the origin of the first hominins can only be verified by biometric protocols. These have been developed with series of current humans and apes. The details of the protocols and statistics of the correlations are therefore published in this technical part of the book. They can be applied in the future to any present-day or fossil species, in order to demonstrate the robustness and efficiency of the method.

5.4.1. *The double craniofacial pantograph*

The relationships between the base, mandible and maxilla during growth are modeled by geometric figures, a triangle for the alveolar arch in the transverse plane and a pantograph in the median sagittal plane, uniting the orientation of the clivus with that of the mandibular symphysis, for example (Figure 5.35).

This model clearly visualizes the difference in amplitude between gorillas and *Homo sapiens* newborns (Figure 5.36).

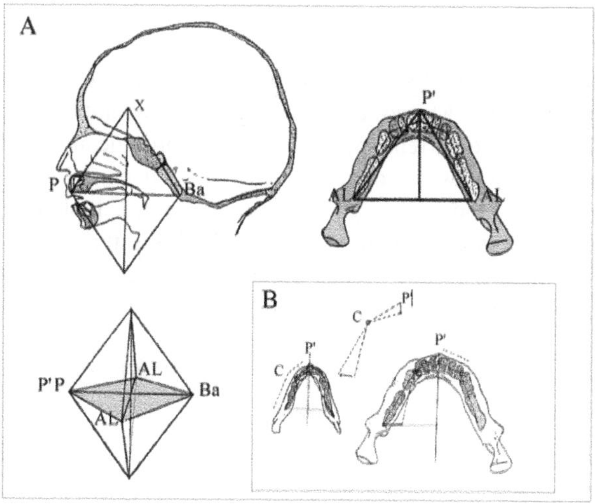

Figure 5.35. *The double pantograph (Dambricourt Malassé 1993): AL: posterior alveolar edge, Ba: basion, C: incisor-canine inter-alveolar point, P: prosthion, P': incisor-canine inter-alveolar point, X: major axis of the clivus passing through the basion and the anterior edge of the sella turcica (ehippion). The major axis of the mandibular symphysis passes through the gnathion and the incisor-canine inter-alveolar point P'. Construction of the pantograph: draw the straight line Ba-P, draw its perpendicular bisector which always passes through the posterior edge of the alveolar arch, draw the slope of the clivus Ba-X to the intersection with this perpendicular, draw the straight line X-P, draw the straight line P passing through the axis of the mandibular symphysis up the intersection, draw Ba up to this intersection*

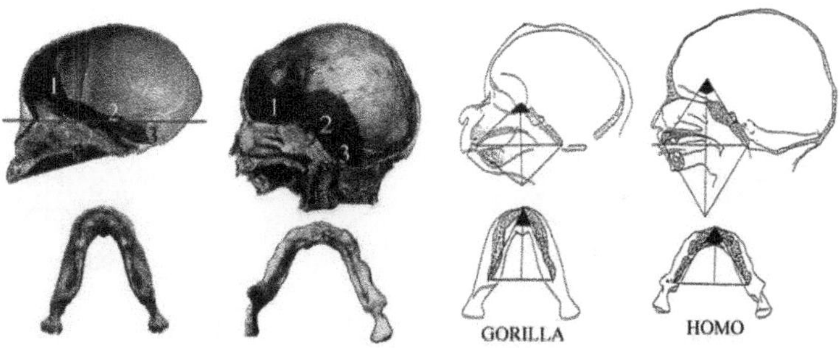

Figure 5.36. *Neonatal stage of a gorilla and* Homo sapiens. *Left: comparison in both planes, sagittal and transverse of the skull base and the mandible. The horizontal is the spheno-ethmoidal plane. The sphenoidal angle is given by its intersection with the axis of the pantograph formed by the clivus (photos: A. Dambricourt Malassé). Right: application of the double pantograph to the two specimens (Dambricourt Malassé 1993). For a color version of this figure, see www.iste.co.uk/dambricourt/embryogeny.zip*

5.4.2. *Internal and external craniofacial contraction angles*

In 1999, the MRI of fossil skulls was still exceptional and access to the internal base was therefore rare. Moreover, studying them without first building up a statistical database for the current species would not provide any further explanation of the differences. Two new protocols have therefore been defined for radiography to add information on the degree of base straightening, looking for correlations between internal and external angular values, in both sagittal and transverse planes, in order to extrapolate them on fossils, either because CT scanning them was not feasible, or because complete cranial bases are rare.

5.4.2.1. *A new angle for a better understanding of the evolution of straightening*

The rotation of the basi-sphenoid being the first movement of the base, it was necessary to quantify it with a new angle called the "angle of sphenoidal rotation". Its adult value is more representative of the cerebro-spinal straightening than the sphenoidal angle, since the latter opens after birth in non-hominid primates. This "sphenoidal rotation" angle has been measured on *Macaca*, *Gorilla*, *Pan* and *Homo sapiens*, and statistics were then presented to the *Société de Biométrie Humaine* (Dambricourt Malassé 2005). The new angle of sphenoidal rotation is measured in the median sagittal plane between two lines traced with the anterior point of the cribriform plate (cr), the anterior clinoid apophysis (ep) and the scar of the spheno-occipital synchondrosis on the pharyngeal surface (s). The protocols were applied to the adult stage of 110 *Homo sapiens* from the IPH multi-ethnic collection, 32 chimpanzees including 6 juveniles and 26 gorillas, 24 macaques including 14 adults (Figure 5.37).

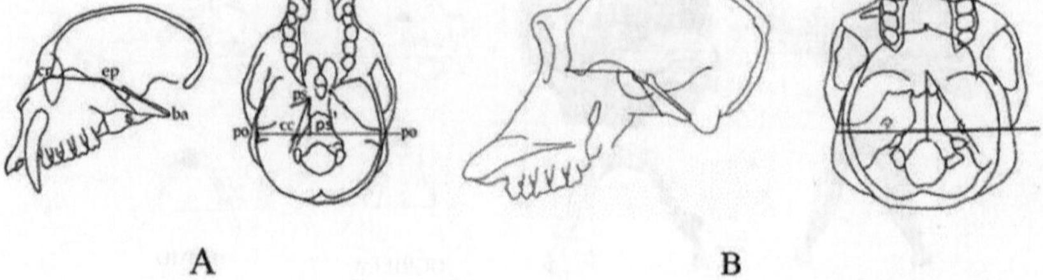

Figure 5.37. *Protocol applied to macaque (Cercopithecoid) and gorilla in lateral and lower views; ba: basion; cc: center of the carotid canal; cr: anterior point of the sieve blade; ep: ephippion; po: porion; ps contact of the anterior end of the petrous pyramids with the body of the sphenoid (Dambricourt Malassé 2005)*

The angles are:

– Sphenoidal rotation angle: cr-ep-s;

– Sphenoidal angle: cr-ep- ba;

– Pyramidal frontalization angle: cc-ps-ps'.

Correlations were calculated using the coefficient of determination r2 (Table 5.3).

Taxons	r2
All species combined	0.63
Anthropoïdea (Homo sapiens, Pan, Gorilla)	0.54
Paninae (Gorilla, Pan)	0.2
Cercopithecidae (Macaca, Semnopithecus)	0.2
Paninae and Cercopithecinae	–0.06
Homo sapiens	0.07
Pan	0.38
Gorilla	0.19

Table 5.3. *Correlation between sphenoidal angle and pyramidal frontalization*

The correlation is maximum for all species combined r2 = 0.63 and is only r2 = 0.54 for *Homo sapiens*, *Pan* and *Gorilla.* Correlations are not observed within the same genus.

Taxons	r2
All species combined	0.72
Anthropoïdea (Homo sapiens, Pan, Gorilla)	0.73
Paninae (Pan, Gorilla)	0.53
Cercopithecidae	–0.12
Paninae and Cercopithecidae	0.4
Homo sapiens	–0.04
Pan	0.4
Gorilla	–0.2

Table 5.4. *Correlation between sphenoidal rotation angle and pyramidal frontalization*

The coefficients of determination are higher for sphenoidal rotation and pyramid frontalization, with 0.72 for all species combined, 0.73 for *Homo sapiens*, *Pan* and *Gorilla*; 0.53 for *Gorilla* and *Pan*; 0.4 for *Paninae* and *Cercopithecus*; and 0.4 for *Pan*.

5.4.2.2. *Conclusion*

The highest coefficients of determination are observed when the three groups, or stages of craniofacial contraction, are combined. The frontalization of the pyramids is correlated to the rotation of the sphenoid and not to the inclination of the clivus, which corresponds to the fact that the central part of the occipital (basi-occipital) follows the trajectory of the vault with the rise of the inion in monkeys and apes. As expected, the new angle is more significant of the embryonic neural dynamics than the classic sphenoidal angle and the cranio-facial contraction is well observed in the two sagittal and transverse planes between two ontogenetic organization planes.

5.4.2.3. *Angles of internal and external craniofacial contraction in the sagittal and transverse planes*

The second protocol aims to identify, on the one hand, what angles are correlated with the sphenoidal angle, and on the other hand, which angles of internal and external contractions correlated (1) with each other and (2) with the angles of transverse frontalization, at the level of the upper face (maxillary) and skull base (petrous pyramids) (Figure 5.38). It is defined as follows:

– in the sagittal plane, the angles correspond to the sphenoidal angle (sph), internal (A) and external (a) craniofacial contraction;

– in the transverse plane, the protocol quantifies the incisor-canine alveolar frontalization (α) as well as the frontalization of the petrous pyramids (β), with a protocol close to that of Dean and Wood (1982).

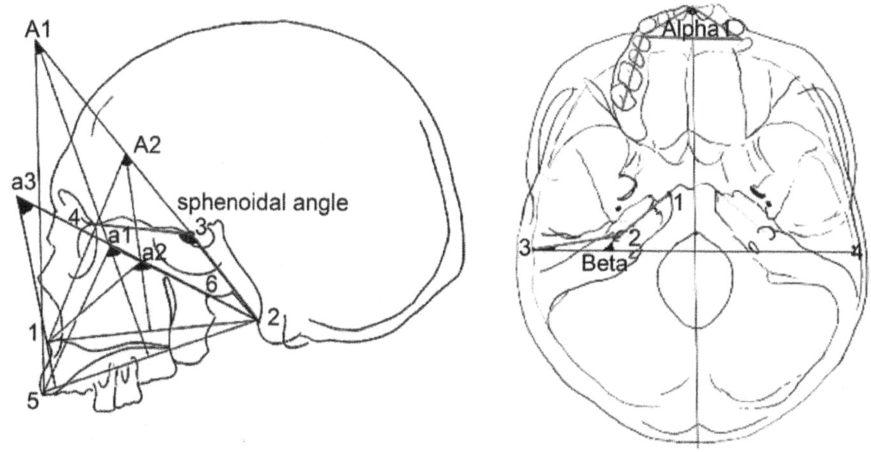

Figure 5.38. *Adult* Homo sapiens, *the skull is oriented according to the ENT standards (orbito-meatal plane). Left: the angles of internal (A) and external (a) contraction; 1: anterior nasal spine, 2: basion, 3: anterior clinoid apophyses, 4: anterior point of the cribriform plate, 5: prosthion, 6: point of the maximum inflection on the external face of the skull base. Right: alpha 1 is the angle of frontalization of the incisor-canine alveoli, measured at the apex of the triangle on the inter-incisal suture, formed by the intersection of two straight lines, each joining the center of the posterior edge of the canine alveolus; beta 1: angle of endocranial frontalization of the petrous pyramids: apex of the triangle formed at the intersection between the straight line joining the anterior end of the petrous bone (1) to the center of the carotid canal (2) extended to the transverse porion-porion axis (3–4) (Dambricourt Malassé* et al. *2000a, 2000b; Dambricourt 2001, 2005). For a color version of this figure, see www.iste.co.uk/dambricourt/embryogeny.zip*

The protocol distinguishes between contraction with alveolar prognathism (prosthion, A1 and a1) and suborbital prognathism (nasal spine A2 or a2 and a3).

5.4.2.3.1. Materials and results

The protocols were applied at the adult stage on 33 *Homo sapiens* and 22 *Paninae*, including 10 chimpanzees and 12 gorillas. The comparison focuses on *Homo sapiens* and chimpanzees. The sphenoidal angle was also measured on six chimpanzees of increasing age from the youngest stage with dm2 (temporary second molar) still included in the alveolar arch. The results are shown in Tables 5.5 and 5.6.

Taxons	**sph**	**σ**	**rsph**	**σ**	**A1**	**σ**	**A2**	**σ**	**a1**	**σ**	**a2**	**σ**	**a3**	**σ**
Sapiens	127	5	119	5	49	8	70	8	83	9	115	9	48	5
Pan	155	6	132	4	88	8	105	8	144	5	162	4	80	2
Gorilla	157	10	137	8	–	–	–	–	–	–	–	–	–	–

Table 5.5. *Angular values and standard deviation of adult contraction angles of* Homo sapiens, Pan *and* Gorilla. *sph: sphenoidal angle; rsph: sphenoidal rotation angle*

Taxons	**A**	**σ**	**β**	**Σ**
Sapiens	26	6	39	6
Pan	45	2	50	4
Gorilla	_	_	70	2

Table 5.6. *Angular values and standard deviation of alveolar (α) and pyramidal (β) frontalizations of* Homo sapiens, Pan *and* Gorilla

Pan	Dental phase	Sphenoidal angle
1974-65	dm2 included	142
1974 33	c included	146
1974-35	M1 included	142
1974-51	M1 in place	157
1974 58	M2 in place	143
1974 29	M2 in place	160

Table 5.7. *Chimpanzee sphenoidal angle by dental phase*

Correlations of the sphenoidal angle (sph)

Homo sapiens in the adult stage:

– with internal angles: A1 r2 = 0.86   A2 r2 = 0.81;

– with external angles: a1 r2 = 0.59   a2 r2 = 0.7   a3 r2 = 0.78;

– with the frontalization of the petrous pyramids: r2 = 0.70;

– with incisor-canine frontalization: r2 = 0.59.

Overall the coefficients are high.

The chimpanzee as an adult:

– with internal angles: A1 r2 = 0.36   A2 r2 = 0.4;

– with external angles: a1 r2 = 0.002   a2 r2 = 0.07   a3 r2 = 0.007;

– with the frontalization of the petrous pyramids: r2 = 0.29;

– with incisor-canine frontalization: r2 = 0.30.

In the adult stage, the coefficients are very low. The sphenoidal angle opens by 20° with growth. The internal contraction is the least influenced by the growth of the face which develops with new functions related to chewing, while the posture moves the occipital away from verticality with the inversion of the growth vectors.

Homo sapiens and the chimpanzee together in the adult stage:

– with internal angles: A1 r2 = 0.8 A2 r2 = 0.8;

– with external angles: a1 r2 = 0.6 a2 r2 = 0.7 a3 r2 = 0.7;

– with the frontalization of the petrous pyramids: r2 = 0.5;

– with incisor-canine frontalization: r2 = 0.48.

The correlations are well observed between the change in amplitude of the sphenoidal angle with the suborbital and maxillary contraction. The correlation with the frontalization of the pyramids and that of the incisor-canine alveoli is more scattered.

Correlations of the sphenoidal rotation angle (rsph)

Homo sapiens in the adult stage:

– with internal angles: A1 r2 = 0.32 A2 r2 = 0.39;

– with external angles: a1 r2 = 0.0 a2 r2 = 0.04 a3 r2 = 0.08;

– with the frontalization of the petrous pyramids: r2 = 0.01;

– with incisor-canine frontalization: r2 = 0.15.

The coefficients are very low. The difference with the sphenoidal angle is explained by the fact that sphenoidal rotation is an early embryonic process, the absence of adult correlations does not signify their absence during the first weeks of development, the development of the different functions has erased them. These losses of correlations can lead to disharmony or even basi-cranio-facial dysmorphosis.

The chimpanzee as an adult:

– with internal angles: A1 r2 = 0.18 A2 r2 = 0.11;

– with external angles: a1 r2 = 0. 07 a2 r2 = 0.04 a3 r2 = 0;

– with the frontalization of the petrous pyramids: r2 = 0.27;

– with incisor-canine frontalization: r2 = 0.12.

The coefficients are very low or even non-existent, because the rotation of the sphenoid is completed very early, the correlations become more blurred with growth than in *Homo sapiens*: (1) the basi-occipital follows the inversion of growth vectors in the occiput, which leads to the opening of the sphenoid angle (bipedal equilibrium less frequent in favor of quadrupedal equilibrium), and (2) the face that develops in front of the upper cerebral fossa (prognathism). The same is true with the frontalization of petrous pyramids.

The chimpanzee and *Homo sapiens* together:

– with internal angles: A1 $r2 = 0.78$ A2 $r2 = 0.89$ A3 $r2 = 0.88$;

– with external angles: a1 $r2 = 0.71$ a2 $r2 = 0.72$;

– with the frontalization of the petrous pyramids: $r2 = 0.39$;

– with incisor-canine frontalization: $r2 = 0.21$.

The change in the amplitude of the sphenoidal angle is well accompanied by internal and external craniofacial contractions and reduction of subnasal prognathism. On the other hand, between the adult stages, there is little correlation with the two incisor-canine and pyramidal frontalizations, as these fade with growth in the chimpanzee, characterized by a large opening of the sphenoidal angle during growth (Table 5.7).

Correlations of internal contraction angles with incisor-canine frontalization:

– *Pan* and *Homo sapiens* together: A1 $r2 = 0.7$ A2 $r2 = 0.62$;

– *Homo sapiens*: A1 $r2 = 0.22$ A2 $r2 = 0.06$;

– *Pan*: A1 $r2 = 0.18$ A2 $r2 = 0.18$.

The correlations are not visible in the adult stage of the same genus; however, they are confirmed with the change in craniofacial contraction between the two genera.

Correlations between external contraction angles

With incisor-canine frontalization (α):

– *Pan* and *Homo sapiens* together: a1 $r2 = 0.72$ a2 $r2 = 0.75$ a3 $r2 = 0.72$;

– *Homo sapiens*: a1 $r2 = 0.17$ a2 $r2 = 0.30$ a3 $r2 = 0.15$;

– *Pan*: a1 $r2 = 0.31$ a2 $r2 = 0.12$ a3 $r2 = 0.23$.

Correlations are confirmed with the change in external contraction (*Pan* and *Homo* together).

With pyramidal frontalization (β):

– *Pan* and *Homo sapiens* together: a1 $r2 = 0.38$ a2 $r2 = 0.35$ a3 $r2 = 0.34$;

– *Homo sapiens*: a1 r2 = 0.037 a2 r2 = 0.006 a3 r2 = 0.00;

– *Pan*: a1 r2 = 0.0 a2 r2 = 0.05 a3 r2 = 0.00.

No correlation is found in the adult stage. Frontalization of the petrous pyramids is not correlated with the degree of external contraction.

Assessment: the increase in internal craniofacial contraction between the chimpanzee and Homo sapiens *is well accompanied by a frontalization of the incisor-canine alveoli and the petrous pyramids. These correlations disappear with the exocranium.*

Correlations between internal and external contraction angles (Table 5.8).

	r 2					
	Chimpanzee		***Homo sapiens***		**Chimpanzee and *Homo sapiens***	
	A1	A2	A1	A2	A1	A2
a1	0.07	0.00	0.71	0.59	0.80	0.74
a2	0.13	0.02	0.87	0.54	0.88	0.83
a3	0.12	0.68	0.88	0.60	0.88	0.85

Table 5.8. *Coefficient of determination between the angles of internal (A) and external (a) contractions in chimpanzees and* Homo sapiens

– *Homo sapiens* and chimpanzee together: the change in craniofacial contraction amplitude shows correlations with all angles.

– *Homo sapiens*: correlations are important with the nasal spine (A1, a1) or suborbital prognathism but decrease with the alveolar end (the prosthion).

– Chimpanzee (*Pan*): the only adult correlation is observed between the internal and external alveolar contraction angles, that is with the prosthion.

5.5. Conclusion: developmental heterochronies and dynamic trajectories

Bolk's theory of fetalization is in line with studies that note the chronological shift in the different stages of cephalic development between large groups of primates, with *Homo sapiens* being the most immature from birth and the latest to reproduce. These shifts are called heterochronies. At the *Muséum national d'Histoire naturelle*, the work undertaken in this direction was continued by Edmond Dechambre (1895–1971), Doctor of Veterinary Medicine, Deputy Director of the *Menagerie of the Jardin des Plantes*, which depended on the chair of wild animal ethology:

> It is in the very intimacy of the nervous system that an embryonic state persists in Man, on which other human characteristics, such as cranial flexion, vertical attitude and all their consequences, would finally depend. Many points are still hypothetical and must be subjected to a new critique before we can measure the value of this theory. (Dechambre 1948–1949, author's translation)

Cranial flexion is not an embryonic persistence, and hominization is not a neotenic process; nevertheless, heterochronies are not new to human paleontology and constitute one of the major approaches for the understanding of the hominization process (e.g. Berge 1998). Stephan Gould (1977) formalized and generalized them to paleontology, but he did not study present-day and fossil primates.

Biometric protocols, or double pantographs applied to taxa representative of large groups of primates, find these discrepancies with the differences in the amplitude of contraction, it being understood that in non-human primates, the basi-occipital (which is lowered before birth) tends to regain the primordial horizontal organization driven by the postural functions of the vertebral axis.

A greater contraction corresponds to a longer trajectory of lowering in space and therefore in duration. The rotations of the basi-sphenoid and the basi-occipital are prolonged between the three great groups: monkeys, apes and *Homo sapiens*. Not only the trajectories, but also the anatomy of the brain and the dentition demonstrate this. All the phases are prolonged from embryogenesis with the disappearance in *Homo sapiens* of the later phases of other primates: the face no longer has prognathic growth, the disassociation of the external table from the frontal bone which forms the torus above the orbits never appears.

The prolongation of the development of neural tube cells can be seen between the great ape and *Homo sapiens*, not only by the contraction of the chondrocranium, but also by the dental germs formed in part by the cells of the neural crests that colonize the facial buds. However, their migration, from the top of the neural tube toward the buds, participates in the tensions of the extra-cellular connective tissues which close the fourth ventricle. Thus, the first temporary molar (dm1) forms only two cusps in the great ape, whereas that of *Homo sapiens* forms five like all the other molars (the three permanent ones and the two temporary ones). In other words, cell divisions have more time to form the five cusps of dm1 in humans, which corroborates a longer closure time of the fourth ventricle. Similarly, the activation of canine growth, which forms a fang in the great ape, or a single cusp that is longer than the others, exceeds their occlusal surface and has disappeared from the temporary dentition in *Homo sapiens*. All these findings, which were presented in 1987 at the IPH and then to the *Académie des Sciences* by Jean Piveteau (Dambricourt Malassé 1988), have been confirmed over 35 years of research. Now, we can observe the embryonic straightening in fossil species and their degree of craniofacial contraction.

6

Fossil Species: From the First Primates to the First Hominids

6.1. Mandibles, witnesses of straightening

Table 6.1 lists the taxa studied in this chapter. The fossils are classified by angle into (1) small catarhinian anthropoids with an adult angular deviation of less than 60°; (2) great apes with the mean angular value of the present ponginae and paninae species; (3) Australopithecine; (4) all *Homo* fossil species combined; and (5) sapiens.

Genus and species	Period	Classification	n	Region
Parapithecus	Late Eocene Lower Oligocene	Small Catarhini	1	Egypt
Apidium	Late Eocene Lower Oligocene	Small Catarhini	1	Egypt
Propliopithecus	Oligocene	Small Catarhini	1	Egypt
Aeolopithecus	Oligocene	Small Catarhini	1	Egypt
Aegyptopithecus	Oligocene	Small Catarhini	1	Egypt
Pliopithecus	Miocene	Small Catarhini	1	France
Oreopithecus	Miocene	Small Catarhini	1	Italy
Limnopithecus	Miocene	Small Catarhini	1	Kenya
Dryopithecus	Miocene	Great Ape	4	Kenya France
Kenyapithecus	Miocene	Great Ape	1	Kenya

Ouranopithecus	Miocene	Great Ape	1	Greece
Sivapithecus	Miocene	Great Ape	5	Indian subcontinent
Gigantopithecus	Pleistocene	Great Ape	4	China
Australopithecus and Paranthropus	Pliocene Lower Pleistocene	Hominidae	29	Africa
Homo habilis, ergaster, sp.	Lower Pleistocene	Hominidae	6	Africa
Homo erectus s.l	Lower Pleistocene	Hominidae	18	Africa Europe Asia
Homo neanderthalensis	Middle Pleistocene Upper Pleistocene	Hominidae	22	Eurasia
Homo sapiens	Upper Paleocene	Hominidae	37	France
Homo sapiens	Epipaleolithic	Hominidae	21	Algeria Morocco
Homo sapiens	Neolithic	Hominidae	57	France

Table 6.1. *Number of mandibles studied by taxon*

Apart from Skhul and Qafzeh, the mandibles of *Homo sapiens* fossils studied come from Europe and date from the Upper Paleolithic (n = 37) and Neolithic period (n = 57), as well as the Epipaleolithic in North Africa (Temara, Taforalt, Morocco, 17,000 years) (n = 21), giving a total of 115 specimens.

6.1.1. *The first primates*

The mammalian species at the origin of the first primate embryogenesis would have appeared between 65 and 55 million years ago (Paleocene). It was classified among the quadrupedal and arboreal Plesiadapiformes with a weight of a few hundred grams. These species lived in the Northern Hemisphere on continents still visible in Europe, Asia and America. Seen from below, the adult skull is clearly divided by a bottleneck behind the orbits between the face and the neurocranium. The palate is very narrow.

As the mode of locomotion is not a factor in distinguishing between a primate and a Plesiadapiform, encephalization is a relevant criterion. It can be seen on the skull of an Adapiform in basal view with the enlargement of the neurocranium behind the orbits and between the pterygoid apophyses, which has the effect of widening the palate and frontalizing the maxillary incisor-canine arch (Figure 6.1(A)). The mandibles of an Adapiform are long and low, forming a narrow V without incisor-canine frontalization (Figure 6.1(B)). The average basilar angular deviation is 34°. The cranial base is flat.

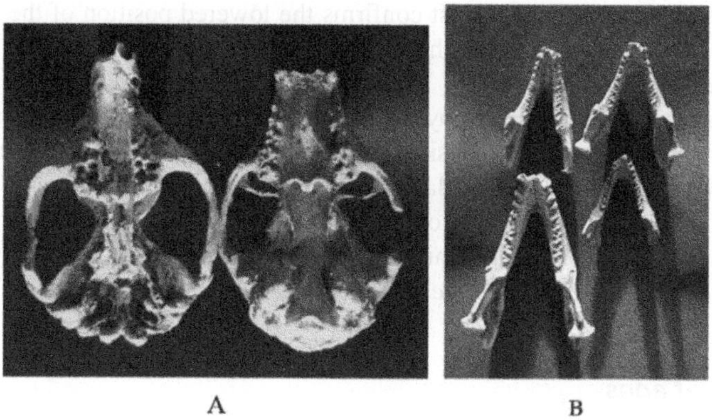

Figure 6.1. *View from below of Plesiadapiform (Adapis) and Adapis mandibles (photo: A. Dambricourt Malassé, different scales). For a color version of this figure, see www.iste.co.uk/dambricourt/embryogeny.zip*

The lineage of prosimian primates at the origin of simian embryogenesis would have been differentiated on continental masses still visible on the Indian subcontinent and in China, distinct from that of the Tarsiers and Strepsirrhines. The axial skeleton and its spheno-basilar straightening (perceptible thanks to the mandible) not being taken into consideration, the criteria are based on the locomotion of the appendicular skeleton, the teeth and the type of tooth wear. Paleoprimatologists are thus divided as to whether or not some fossils are Adapiformes or Simiiformes. These Asian fossils are classified in two families, *Amphipithecidae* and *Eosimiidae*.

6.1.2. *The first simians or monkeys*

In 1987, the oldest mandibles attributed to Simiiformes were collected in Egypt in the Fayum, now known to be 30 million years old (Oligocene), along with *Parapithecidae*, which are believed to be derived from a species of *Eosimiidae* (Eocene). The basilar angular deviation of *Apidium* and *Parapithecus* varies from 60° to 57° like for a small monkey, but the arch is V-shaped, like an Adapiform.

Due to the absence of a skull base, the data is insufficient for a decision to be made. However, other mandibles classified in the family *Propliopithecidae*, dated from 32 to 29 million years ago, are true catarhinian simians; their dental pattern is visible in gibbons, great apes and *Homo sapiens*. Their shape corresponds to the present-day monkey: U-shaped but narrow, with an angular deviation of 57° and a symphyseal angle of 120°, much more upright than an Adapiform. Subsequently, a few well-preserved skulls of the genus *Aegyptopithecus* were scanned (Elwyn *et al.* 2007). The virtual endocranial cast confirms the lowered position of the cerebellum relative to the brain. The basi-sphenoid is thus tilted downward as in the current *Hylobatidae* (gibbon, siamang). Thus, for the first time in the evolution of vertebrates, the dorsal cord rose by a few degrees between 60 and 35 million years ago, most probably within an Asian "prosimian" species. The later mandibles of *Limnopithecus* (Kenya) (20–16 Ma) have a basilar angular deviation of 56°, much like the *Oreopithecus* (Italy) of 8 Ma. They are monkeys distinct from *Cercopithecidae* by their teeth, which could be classified in a superfamily including *Propliopithecidae* and *Hylobatidae*.

6.1.3. *Great apes*

6.1.3.1. *The first fossils, Africa, 20 Ma*

The oldest mandibles attesting to the amplification of the contraction of a species of monkey were collected in Kenya, in Miocene geological formations dated to 20 million years ago; they are related to the genera *Dryopithecus*, *Proconsul* and *Kenyapithecus*. These are great apes (Figure 6.2). In accordance with the predictions inferred from the mandibles of Asian ponginae, African paninae and Lartet's European *Dryopithecus*, fossils of this "ontogenetic morphotype" are found scattered between Africa and Eurasia. They are similar in their dentition to *Propliopithecidae*, but their angles are typically those of the paninae and ponginae; they all have a canine tooth higher than other dental crowns and a first canine-like premolar, with the wear facet formed by the friction of the upper canine tooth. The canine lock attests to a semi-erect occlusal-postural balance, regardless of the locomotor specialization of the appendicular skeleton. African fossil localities are the oldest and are circumscribed to Kenya and bordering Uganda. The oldest fossil is a fragment of a 24 Ma maxilla named *Kamoyapithecus*. An incomplete skeleton is attributed to the genus *Dryopithecus nyanzae* (later *Proconsul nyanzae*). The species is small in size: the base of the mandible measures 56 mm. The animal must have weighed 14 kg. The authors who studied it conclude:

> Proconsul was more encephalized than the modern monkey of comparative size, we suspect that it is a pongid trait. (Walker *et al.*, *The skull of Proconsul africanus*, 1983)

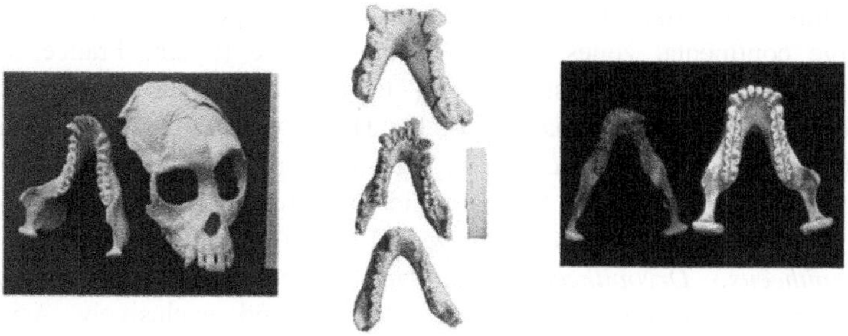

A B C

Figure 6.2. *Comparison of fossilized great ape mandibular shapes (Rusinga Island in Lake Victoria) and present-day apes. A: Cast of the incomplete skull of* Dryopithecus nyanzae *with its mandible, B: three mandibles KNM RU 2087, KNM RU 2036 A and KNM RU 1855, C:* Dryopithecus africanus *and a gorilla (photos: A. Dambricourt Malassé, different scales). For a color version of this figure, see www.iste.co.uk/dambricourt/embryogeny.zip*

The endocranial casting was quite similar to that of the *Hylobatidae* and *Cercopithecus*; however, the base of the neurocranium and the occipital squamous are missing. The 1983 skull reconstruction deprives the fossil of its cerebellar fossa and occiput. Behind the mastoid process, the space of the neurocranium corresponding to the occipital lobe of the cerebellum is missing. The cranial capacity evaluated at 167 ml and the encephalization coefficient at 1.5 are thus underestimated. The occipital and cerebellar volumes need to be taken into account, and imply a much higher encephalization coefficient. It is therefore a great ape, with encephalization as the main reason, because the mandibular basilar angular deviation is 67° and the symphysis angle is 120°. The canine lock is typical of a great ape. The *Proconsul nyanzae* fits into the embryonic neural dynamics common to paninae and ponginae, with a more pronounced brain stem straightening than that of a *Propliopithecidae* or *Cercopithecidae*. At birth, the cerebellar fossa was as low as that of a chimpanzee or a gorilla. The appendicular skeleton attests to an arboreal and quadruped locomotor mode, but its axial occluso-postural balance also allowed for the occasional use of bipedalism.

The middle level of the face is also known for *Dryopithecus major* and *Ugandapithecus major*. Neither shows the tilting of the spheno-ethmoid-facial block of the orangutan. The chondrocranial morphogenesis of the African great apes represents the "ancestral pattern", also called "plesiomorphic" by cladists, within the limits of which the Asian lineage of the orangutan became differentiated during intrauterine development.

In Eurasia, the fossils are more recent, at 13–6 Ma, perhaps 5 Ma, distributed in two large continental zones, from Western Europe (Spain, France, Germany, Hungary and Greece) to Western Asia (Turkey, Georgia and Iran), then, after a fossiliferous gap, from the margins of the Himalayas to the Indian coast of the Arabian Sea (Bhandari *et al.* 2018), then to Southeast Asia, passing through Myanmar (Burma), Thailand, Vietnam and southern China. For the first Eurasian zone, the genera are *Pierolapithecus* (13 Ma), *Anoiapithecus* (11.9 Ma), *Hispanopithecus*, *Dryopithecus*, *Danuvius*, *Graecopithecus*, *Ouranopithecus*, *Rudapithecus*, *Ankarapithecus* and, for the second exclusively Asian zone, *Sivapithecus*, *Indopithecus*, *Khoratpithecus* and *Lufengpithecus*.

All Eurasian species disappeared after the Miocene, or even, for the last ones, during the Pliocene, with the exception of the *Gigantopithecus*. This genus, with a particularly massive head, appeared around 2 Ma in China and Southeast Asia and is said to have been extinct for 100,000 years. The only genus still alive in the lineage of *Sivapithecus-Khoratpithecus* is the island orangutan of Sumatra and Borneo.

The basilar angular deviation greater than 60° is observed on the 15 great ape mandibles (Table 6.1) with a mean of 70° ± 5; the symphysis is sagittal and median with a mean angle of 127° ± 7. The tilting of the spheno-ethmoidal block of the orangutan can be observed from West Asia to southern China with *Ankarapithecus* in Turkey, *Sivapithecus sivalensis* on the Indian subcontinent and a genus wrongly classified in *Lufengpithecus* in southern China. But other genera have not developed this morphogenetic divergence, such as *Anoiapithecus* and *Pierolapithecus* in Europe and *Lufengpithecus* in China.

There are only three complete skulls of great ape fossils, dated between 7 and 6 million years ago; two *Lufengpithecus* and the *Sahelanthropus*, which was nicknamed "Toumaï" and discovered in Chad in 2002. The latter is the only usable one. This fossil has caused considerable controversy – great ape or hominin? – because of the position of the occipital hole studied exclusively from the external view. Until recently, paleontologists had no curiosity for the sphenoid and the angles of the internal straightening of the skull base.

The great apes have thus populated Eurasia by expanding their ecological niche from Africa. Fossils are collected in tectonic zones that favor the upwelling of the deepest sediments. Thus, necessarily huge and apparently barren geographical territories, such as the Arabian Plate and the Indian subcontinent, have been populated, which represents a significant potential for paradigmatic "revolutions" as Miocene and Pliocene fossil pockets are discovered or more actively prospected. A recent discovery in India has disrupted specifically the range of the great apes of the Indian subcontinent and southern China, which were thought to be confined to the landforms marking the tectonic collision between the Indian and Asian plates. In

2011, a jawbone of Miocene age similar to *Sivapithecus* was collected toward the coast of the Arabian Sea, in the state of Gujrat, bringing the range closer to the Arabian Plate (Bandhari *et al.* ibid.).

The Asian great apes had finally lost the anatomical "arguments" which made it possible to consider Asia as a hotbed for the emergence of Hominids, in the sense of the neurax verticalization from the sphenoid. However, the great apes all have the same axial endoskeletal straightening and it is the accentuation of this embryonic straightening that generates the verticalized Hominidae. The straightening is not seen gradually on the appendicular skeleton, and as Teilhard de Chardin predicted, fossil bones always appear too specialized to be an ancestor of the Hominidae. As this shared ancestry cannot be read in arboreal locomotion or the quadrupedal mode of locomotion on the ground, it is appropriate to consider the appendicular pre-adaptation of certain great apes. And this is how certain Chinese fossils pose questions regarding the absence of arguments which envisage a center of hominization, or embryonic straightening, in Asia.

6.1.3.2. *Asia: a great bipedal monkey of 6 Ma*

In southern China, more precisely in Yunnan, different species of great apes reproduced for at least 8.2 million years, until around 1 million years ago, in the form of the *Gigantopithecus*. Yunnan and the surrounding areas are the only region in the world where primatologists can follow the evolution of great apes over such a long period (Yue *et al.* 2004):

> Late Miocene and Pliocene hominoids from Yunnan Province in southern China have been recovered from four sites or site complexes: Xiaolongtan, Yangyi, Shihuiba and Yuanmou. Of these, Shihuiba and Yuanmou are among the most prolific fossil hominoid sites in Eurasia, and they have yielded important evidence that is critical for documenting the evolutionary history, biogeography and paleobiology of later Neogene hominids. (Harrisona *et al.*, *On the systematic status of the late Neogene hominoids from Yunnan*, 2002)

Within this vast diversity of great apes, one species stands out, exceptionally well illustrated by elements of the post-cranial skeleton, such as the shoulder, phalanxes and a very rare proximal end of the femur, which attests to a locomotor mode that was frequently bipedal: *Lufengpithecus lufengensis*, discovered in 1975 in Shihuiba, near Lufeng, and dated at 6.2 Ma. The Shihuiba fossils were collected in an ancient swamp on sedimentary deposits of little more than 2 m. The collection includes five skulls, two of which are complete, 47 cranial fragments, 10 mandibles and 650 individual teeth. The anterior appendicular skeleton is illustrated by a scapula, a clavicle, a radius (bone of the forearm) and three phalanxes of the hand, the posterior skeleton identified thanks to the left femoral extremity and a metatarsal.

We were able to observe the fossils with Ji Xueping at the Institute of Cultural Relics and Archaeology of Yunnan in Kunming and with Zhao Lingxia at the Institute of Vertebrate Paleontology and Paleoanthropology (IVPP) in Beijing. The mandible is about 15 cm long, and the complete set of teeth with the canine lock is typically that of a great ape. The study of tooth eruption showed a growth rate comparable to that of *paninae* and *ponginae* (Zhao *et al.* 2008). Both skulls are complete but excessively compressed, and the three-dimensional reconstruction of the base remains particularly complicated.

Paleontologists Xu and Lu (2008) have published interpretations: the fingers of the hands are adapted to climbing, the shoulder is close to that of an orangutan, the radius resembles that of a chimpanzee and the metatarsus resembles that of a gorilla, and corresponds to its style of walking. On the other hand, the incomplete femur has retained the articular condyle and a segment of the diaphysis and does not resemble any of the three current genera, neither Lartet's *Dryopithecus fontani* (France), *Australopithecus* or *Homo habilis* from Kenya (KNM-ER 1481), nor *Homo sapiens* (Figure 6.3).

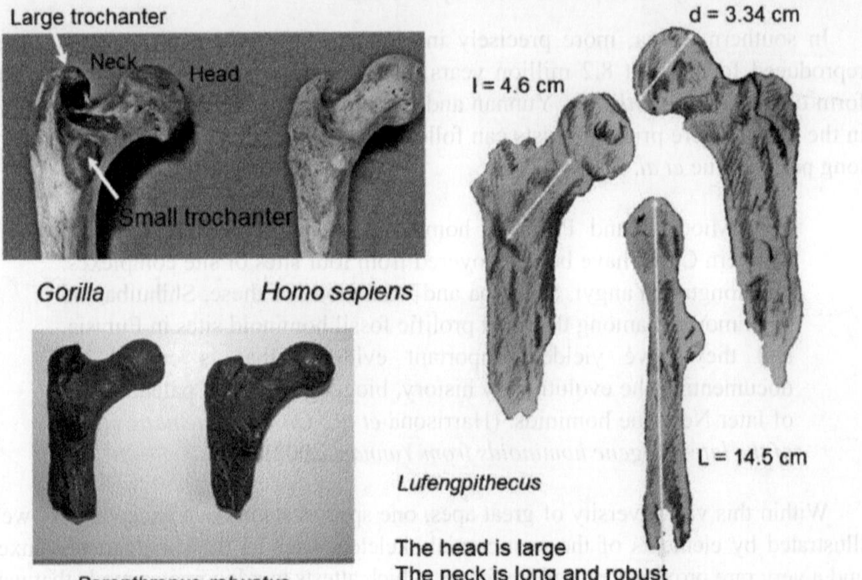

Figure 6.3. *Comparison of the proximal extremities of the femur of Gorilla,* Homo sapiens, Paranthropus *(photos: A. Dambricourt Malassé) and* Lufengpithecus *(drawings: A. Dambricourt Malassé from Xu and Lu (2008), measurements added). The scales are different. For a color version of this figure, see www.iste.co.uk/dambricourt/embryogeny.zip*

Its conformation suggests a more vertical body load than in current great apes. The location of the muscle insertion reliefs allowed their functions to be extrapolated: the thigh was in a flexed position, its muscles and those of the lower leg allowed movements of a small amplitude, and not strides, as in humans. Xu and Lu concluded:

> The *Lufengpithecus* femur could have possessed somewhat strong femoral abductors and femoral adductors. The musculature implies that the trunk and the leg of *Lufengpithecus* possessed preliminary upright posture. And the *Lufengpithecus* femur could have possessed somewhat developed femoral extensors and femoral flexors and the ability to walk bipedally, but only with the leg slightly bent and a markedly sharp step. (Xu and Lu, *Lufengpithecus lufengensis, an Early Member of Hominidae*, 2008).

As for the middle level of the face, it also deserves attention. It is neither the face of the orangutan nor of the *Sivapithecus sivalensis*: the upward tilting of the maxilla is not observed, the orbits are round and not oval, the space between the two orbits is very wide, the zygomatic arch is high and it reaches the great transverse axis of the orbits (Figure 6.4.). The middle level of the face resembles its contemporary, the 5.8-Ma-old *Ardipithecus* of Ethiopia with its round orbits, and the *Paranthropus* with the high position of its zygomatic arch. More recently, a skull from a young 6.1 Ma specimen was collected at Shuitangba, North Yunnan. Its orbits are round and its face is not tilted (Ji *et al.* 2013).

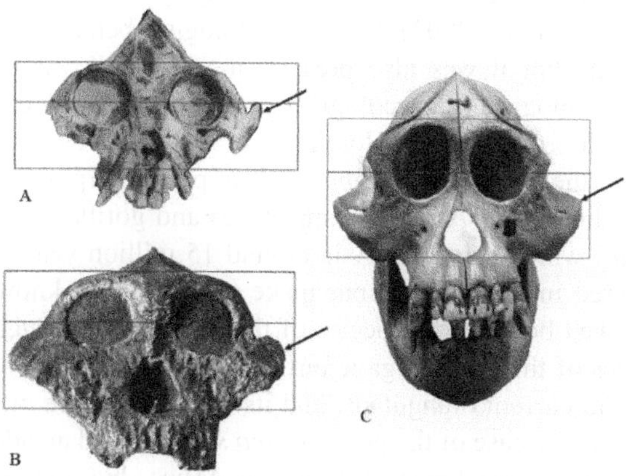

Figure 6.4. *Position of the zygomatic arch relative to the center of the orbits in* Lufengpithecus *(A),* Paranthropus robustus *(B) and* Pongo *(C). (A: drawing: A. Dambricourt Malassé from Xu and Lu (2008), B and C: pictures: A. Dambricourt Malassé, the scales are different). For a color version of this figure, see www.iste.co.uk/dambricourt/embryogeny.zip*

Pongo's lineage lived in this region between 8.2 and 7.1 million years ago, as shown on the middle face of a young specimen found in 1988 in Yuanmou. Observed in Kunming, the fossil shows the upward tilting of the maxilla and the two oval-shaped orbits, so it is not the genus *Lufengpithecus*, but another genus of *Pongo's* lineage that could be renamed *Yunnanpithecus hudienensis* and not *Lufengpithecus hudienensis*.

Lufengpithecus lufengensis is therefore particularly instructive. The occluso-postural balance is that of a great ape, its axial skeleton was thus semi-erect from the basi-sphenoid to the sacrum, it was arboreal and quadrupedal, but the posterior appendicular skeleton reveals a terrestrial displacement in a frequently upright posture. *Lufengpithecus* breaks the paradigm of the single lineage *Sivapithecus-Khoratpithecus-Gigantopithecus-Pongo*, at risk of extinction due to an overly arboreal locomotion and unsuited to the terrestrial way of life, unless it survives in endemic forest environments, such as the orangutan. This aptitude for bipedalism does not make it a *Hominidae*, but it no longer allows the argument of arboreal specialization to exclude the possibility of a hotbed of hominization in tropical Asia.

If infants were born with a verticalized axial endoskeleton no longer permitting quadrupedics – the skeleton covers the central nervous system – the frequently bipedal terrestrial lifestyle of mothers would have favored the reproduction of these "mutants".

This pre-adaptation existed at the same period, around 6 million years ago, in Africa, as evidenced by the femur of the genus *Orrorin*, discovered in Kenya in the Tugen Hills (Pickford *et al.* 2002). Paleoprimatologists believed it was exclusive to the African tropics, but it was also present in Asia. Following the reasoning of Lamarckian paleontologists, the ecological pressure of global cooling could have therefore favored the emergence of Hominids in South Asia as well. Let us push this reasoning further: the face of *Lufengpithecus* is plesiomorphic, relative to that of *Sivapithecus*; the lineage of African chimpanzees and gorillas is thus closer to that of *Lufengpithecus*, which spread in Asia around 15 million years ago. The isolated jawbone discovered in Gujrat does not make it possible to know if the face was tilted: it may as well belong to a species of the lineage of *Lufengpithecus*. In other words, if a species of this lineage gave birth to a *Hominidae*, it would not have the genes derived from current orangutans, and its DNA would be closer to that of the chimpanzee. This is the case of the genus *Homo sapiens*, and an additional argument for locating the emergence focus of *Homo* in Africa, but southern India and the Arabian Peninsula may hold surprises in the fossiliferous formations of the Mio-Pliocene transition, between 5 and 4 million years ago.

6.1.3.3. *The only complete skull, Sahelanthropus tchadensis, 7 Ma, is a great ape*

This 7 Ma fossil is exceptional because it is the only complete fossil skull of a large anthropoid before the appearance of the hominins. Erosion has destroyed some of it and the weight of sediment has deformed the fossil, but the structure is identifiable.

The endocranial base has not been reconstructed, even though this is the determining criterion for deciding whether the specimen is hominian (upright brain stem) or simian (semi-upright brain stem). Its hominian interpretation (Brunet *et al.* 2002) has given rise to very deep disagreements that are still relevant today, insofar as paleontologists know they are confronted with the mechanisms of hominization. Wolpoff *et al.* (2002, 2006) have opposed the hominian interpretation of the fossil.

The question is not to know if Toumai practiced bipedalism, but to be able to verify the embryonic contraction of the fourth ventricle and thus verify if this species passed the threshold of the great apes, if the petrous pyramids and the clivus reached endoskeletal verticality. However, the fossil does not present any characteristics expected from this embryonic threshold.

A mandibular body retained the symphysis, a canine, premolars and molars with their cusps. The shape of the alveolar arch is upsiloid, the radiography does not reveal a symphyseal void, the canine is stocky and protrudes beyond the occlusal surface of the molar rows. Observation of the deformed cast of the skull of *Sahelanthropus tchadensis* shows that the canine tooth was not small, it was split and broken: it was long, voluminous and protruded beyond the occlusal surface of the still visible molars, its root was particularly developed. The occlusion thus formed a canine lock. The prognathism was not reduced either, in basal view; this impression is due to the very clear disappearance of the incisor alveolar arch. The sub-nasal prognathism was destroyed (Figures 6.5 and 6.6).

This skull is thus integrated into the craniofacial morphogenesis of the great ape. The slope of the clivus could not be more straightened than in the neonatal stage of the chimpanzee or gorilla. However, one of the arguments put forward to classify it as a *Hominidae* is the position of the occipital hole. It is said to be more advanced at the bottom. The base underwent a very strong postmortem deformation that completely deviated its antero-posterior axis sideways and moved the occipital hole forward (Figure 6.6).

Figure 6.5. *Casting of the deformed skull of* Sahelanthropus *oriented in the Frankfurt plane. The canine tooth, reconstituted with modeling clay, protrudes beyond the occlusal plane of the other teeth (photo: A. Dambricourt Malassé). For a color version of this figure, see www.iste.co.uk/dambricourt/embryogeny.zip*

Figure 6.6. *Comparison in lower view of the cast of the deformed skull of* Sahelanthropus *and an adult gorilla. The transverse straight line joins the extreme edges of the temporomandibular joint, the short arrow indicates the basion, deflected forward and swung to the side, the long arrow indicates the disappearance of the incisor-canine arch (photo: A. Dambricourt Malassé, same scale). For a color version of this figure, see www.iste.co.uk/dambricourt/embryogeny.zip*

The occlusion already predicted anomalies in the repositioning of the occipital hole. Indeed, a comparison between the physical cast and the virtual reconstruction of the exocranium highlights the disappearance of the external auditory canal, causing the alignment of the basion on the Glaser fissure (tympano-squamous suture), which generates an anatomo-physiological nonsense. The basion is advanced by one centimeter, which can only considerably modify the unknown angular values of the sphenoidal angle (Dambricourt Malassé 2006a, 2006b, 2009, 2010, 2011a). The basion was therefore necessarily in a more posterior position and the occipital hole was less advanced.

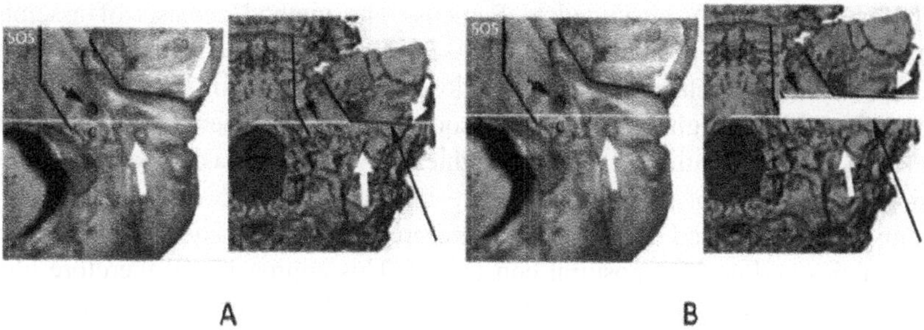

Figure 6.7. *Comparison of the positions of the auditory canal, petrous pyramid and temporomandibular joint of the gorilla with the virtual reconstruction of* Sahelanthropus *by ZolliKofer (Brunet et al. 2002). A: before and B: after correction. C and small black arrow: jugular foramen; white arrow below: stylo-mastoid foramen; white arrow above: Glaser's fissure; long black arrow: abnormal alignment of the scissure on the tangent to the basion; SOS: spheno-occipital synchondrosis. B: shift of the Glaser's fissure to leave the auditory canal space missing in A*

Figure 6.7 shows how, in the gorilla, the tangent to the basion (white straight line) cuts the auditory canal and passes below the carotid canal (small black arrow) and the Glaser fissure (white arrow below). The fissure is at the level of an angle bordering the clivus. Detailed observation shows the disappearance of the auditory canal on the virtual reconstruction and alignment of the basion on the carotid canal and the Glaser fissure (long black arrow). An offset of the fissure is necessary to reconfigure the topography of the petrous bone in a much more legible and coherent way. As a result, the carotid foramen is in front of the basion, which moves further back, lengthening the pharyngeal face of the clivus. The virtual reconstruction has thus moved the occipital hole forward, forcing the removal of the auditory canal and generating a conformation which does not exist in primates. The base was therefore that of a great ape. Nothing suggests a relationship with Hominids; this skull may also be in the lineage of *paninea*. The basi-cranio-facial conformation in extension and the occlusion are consistent with a great ape-like femur, still unheard of, according to Roberto Macchiarelli and Aude Bergeret (Callaway 2018).

6.2. The great ape-Hominidae transition: an acceleration in the complexity of embryogenesis

6.2.1. *Ardipithecus: 5.8 to 3.8 Ma, Ethiopia*

The example of the *Sahelanthropus* skull illustrates the postulate of gradualism imposed on the reconstruction of fossils and their naming, which will be repeated with fossils arbitrarily attributed to Sapiens. The method consists of assembling fossils following the postulate of the conservation of great ape characteristics and the appearance of characteristics as close as possible to *Homo sapiens*, not concerning axial straightening, but the acquisition of permanent bipedalism and the reduction of prognathism. The straightening of the basi-sphenoid and the basi-occipital are not a problem taken into account between great apes and Hominids. The face and occlusion are therefore studied independently of sphenoidal biodynamics and occluso-postural constraints. This approach will therefore not use the expected knowledge to situate the current disorders of occluso-postural balance in the curve of supra-chordal axial straightening, which is still not perceived as an evolutionary problem.

Like many extinct genera, *Ardipithecus* is composed of fossils from different localities and geological strata of varying ages. Yohannes Haile-Selassie has been collecting fossils of the oldest species in Afar since 1999, dated from 5.77 to 5.54 Ma, which lived in an environment similar to the *Sahelanthropus*, consisting of forests, wooded savannahs and areas of open water. The species is illustrated by an incomplete mandible, upper and lower canines, hand and foot phalanxes, and an arm bone. It was named *kadabba* in 2004 (Haile-Selassie *et al.* 2004). The symphysis is incomplete, but the body of the mandible is sufficiently preserved to be restored symmetrically.

More recent fossils had been collected five years earlier, in 1994, on the shores of two tributaries of the Awash River, Adgantoli and Aramis, over a distance of 2 km, at five localities and in geological strata between 4.4 and 3.8 Ma (Haileab and Brown 1992; Woldegabriel *et al.* 1994). The base of an *Australopithecus* skull of 3.9–3.8 Ma (ARA-VP-1/500) was found more than 500 m away from a maxilla, cranial vault and iliac wing collected in layers dating from 4.4 Ma (ARA-VP-6/500). Stratigraphic logs show more than 2 m of sedimentary deposits between the two layers. All of these fossils have been associated. The reconstructed skull and face are of the type of a great ape with an upsiloid arch and a canine lock (Figure 6.8). The arbitrary association of the Australopithecine base with these simian fossils prompted paleontologists to suggest the name *Australopithecus ramidus* (White *et al.* 1994). Then, facial features prevailed and a new genus *Ardipithecus*, from the species *ramidus* was created (White *et al.* 1994). Fossils from the different localities were published in detail 15 years later by Science (2009).

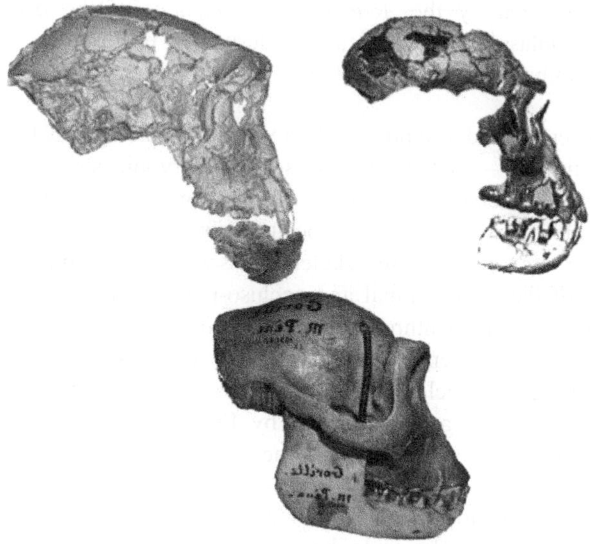

Figure 6.8. *Comparison of* Sahelanthropus *and* Ardipithecus *occlusions after replacing a mandible of the taxon in dental articulations, with that of the gorilla. The temporal of* Ardipithecus *associated with the petrous pyramids was erased, as it is an* Australopithecus *(ARA-VP-1/500) from a more recent site (Dambricourt Malassé 2010) (photo of gorilla; A. Dambricourt Malassé,* Sahelanthropus *according to Brunet* et al. *(2002, 2005);* Ardipithecus *according to Suwa* et al. *(2009). The scales are different). For a color version of this figure, see www.iste.co.uk/dambricourt/ embryogeny.zip*

As seen with the temporary dentition of the great apes, it is not so much the size of the crowns that is important as the intercuspidation of the two dental arches with the canines, the first molar and the lateral incisor. *Ardipithecus*' occlusion is no less locked than the temporary one of the chimpanzee, or *Sahelanthropus*. The upsiloid shape of the arcade indicates the absence of lateral encephalization of the anterior end of the temporal lobes. The face and mandible are typical of a semi-erect occluso-postural balance.

Yet, the cerebellar fossa and the clivus of an *Australopithecus* were verticalized with a very pronounced contraction of the basi-sphenoid, which was found in the mandible with a more upright symphysis, the appearance of the symphyseal void, the disappearance of the canine lock and a more flared molar arch with an incisor-canine frontalization. The association of the base with the face of an ape is therefore an embryological counter-sense, making the composite skeleton of *Ardipithecus ramidus* a veritable chimera, reminiscent of the myth of the missing link. The contradictions were demonstrated in 2004, during the international colloquium organized by the *Académie des sciences* of the *Institut de France* (Dambricourt Malassé 2006a), detailed during a posturology and osteopathy colloquium (Dambricourt Malassé 2006b).

The geological period around 4 million years ago corresponds to the period from which the Hominids – in the strict sense of embryonic straightening – reached demography and dispersion in Africa sufficient for *Australopithecus* fossils to be statistically collected in the territories occupied by the *Ardipithecus*. It is therefore imperative not to presuppose skeletal associations, remembering (1) the understanding of the cranio-spinal and occluso-postural morphogenetic unit is not within our ability in paleoanthropology and (2) partnerships with orthopedists and occlusodontics are indispensable. The doctrine of mosaic gradualism does not conceive the phases of acceleration in the generalized morphological changes from the head to the sacrum, as summarized by Etler: "To go from *Ardipithecus* to *Australopithecus* in such a short span of time seems hard to imagine" (Etler 2010).

The rapidity of transformations with angular discontinuities is possible thanks to homeotic genes which act alongside the embryonic axis. It is not necessary to create missing links with mixed great ape and human straightening characteristics. The fossils will be great apes or hominids, but certainly not intermediaries like Dubois' *Pithecanthropus*, with a very low cranial vault and a *Homo sapiens* femur.

In 2016, a complete *Australopithecus* skull was finally discovered in Ethiopia at Woranso-Mille, dated 3.8 Ma, MRD specimen VP-1/1 (Haile-Selassie *et al.* 2019). No paleontologist has classified it in *Ardipithecus*, although its base is identical to ARA-VP-1/500, confirming the chimeric nature of the composite skeleton. The canine teeth show very pronounced end-to-end wear, which is indicative of a change in occlusion and cranio-spinal balance. The canine lock has disappeared in relation to the disappearance of the equilibrium constraints of the skull cantilevered over the spine. This change corresponds to an evolution in the embryogenesis of neural ridges which also prolongs the morphogenesis of dental germs and makes the canine, temporary and first premolars more delicate.

A second skull of *Australopithecus prometheus* in South Africa, almost as old, at 3.67 Ma, completes the knowledge of the endocranial base with the famous Little Foot discovered by Ronald J. Clarke at Sterkfontein (Stw 573) (Clarke and Kuman 2018). For the first time, a complete *Australopithecus* skull with its mandible was associated with the spine and the appendicular skeleton. The two canines are end-to-end and show apical wear, like the Ethiopian skull of Woranso-Mille. The base is complete but the lower view shows a post-mortem displacement of the clivus anterior to the ear canal, which has the effect of aligning the basion with the Glaser fissure.

6.2.2. *Where to find their origins: trees, tall grasses or the placenta?*

Locomotor and masticatory functions can no longer be the only keys to reading morphological changes, especially since this Lamarckian way of thinking avoids the question of the heredity of acquired traits. It is therefore advisable to remain cautious when it comes to skeletal reconstructions from isolated bones which come from different geochronological horizons, because great ape embryogenesis such as *Ardipithecus* and that of hominids emerging from it necessarily co-existed in the same tribes. It took a number of generations to stabilize the reorganization of populations progressively split into apes and hominids, given the high rate of abortions due to the incompatibility of embryonic memories transmitted by female and male gametes of the old and new organizational plan.

Axial verticality appeared in amniotic fluid as a result of the complexification of embryogenesis and the regulation of homeotic genes. Complexification can only be generated during gametogenesis, and regulation is the result of properties transmitted over millions of years by meiosis. Our conclusions have always placed the origin of arboreal hominids in the environment of the predilection of pregnant and parturient females in search of water for amniotic fluid and lactation, and mastering the conditions of survival and safety of the young for many generations of great apes. These environments were humid and tree-covered and have certainly remained the preferred living conditions for hominids. Behavioral adaptations to more open environments were later supplemented by technological innovations necessary for developing subsistence strategies and defense against predators.

We have never subscribed to the idea that the climate forced the body to straighten, as Phillip Tobias agreed, considering wetlands such as rivers, which are preferred travel corridors. South African karsts are aquifers, territories where water circulates in the soil and appears on the surface when the water tables rise due to erosion. Recent palynological studies of the habitat areas of *Ardipithecus ramidus* in the Awash (Ethiopia) and Lake Eyasi (Tanzania) have confirmed the relevance of our findings, disseminated for more than 30 years. The results show that arid East Africa offered a wide variety of micro-habitats with wetlands which remained densely forested thanks to groundwater, such as palm forests (Barboni *et al.* 2019).

It is therefore logical to find more fossils of arboreal great apes like *Ardipithecus* from 5.5 to 4.5 Ma in the same wooded territories, while the density of hominins populations, still arboreal, was statistically too low to leave fossils. These critical phases of tribal reorganizations are at the heart of the processes of the hominization of the body and the humanization of conscious behaviors in the psychomotor learning period.

Fossil mandibles and the preserved rare skull bases confirm not only the existence of a united organizational plan of embryonic origin, but also the fact that this morphogenetic unit, or more precisely, its molecular and dynamic memory, is the ancestor of the organization of *Homo sapiens*.

6.2.3. *Australopithecus* and the oldest species that defines it: *anamensis* or *afarensis*?

The species *anamensis* was defined in 1996 (Ward *et al.* 1999), with fossils collected in 1994 and 1995 during surveys covering the same geological period as in Ethiopia, but this time in Kenya, on the shores of Lake Turkana (East and West Turkana). The fossils come from two major localities: Kanapoi, located in the southwest, dated 4.2 Ma, where a humerus of very human conformation had been discovered in 1965; and Allia Bay, dated 3.9 Ma, located northeast of the lake. The example of *Ardipithecus* shows that one should be cautious with regard to assemblages for this geological period. The collection totals 21 specimens, 12 at Allia Bay and nine at Kanapoi with the humerus, a mandible, a palate, the distal and proximal extremities of an almost complete tibia, and cranial fragments.

The Kanapoi mandible (KNM-KP 29281) has the same characteristics as *Ardipithecus* with an upsiloid arch, to which an imposing alveolar planum that accentuates the fleeing profile of the symphysis is added. The planum is longer than in *Australopithecus*, the insertions of the genio-glossal muscles of the tongue are of simian type and not hominian. The canines are robust with a large root, the proximal cusp of the first premolar is developed and corresponds to a canine morphogenesis of a great ape with a wear facet. The canine is a hook held in its alveolus by a powerful root and shows beveled wear rather than end-to-end wear. The narrow U-shaped palate arch (KNM-KP 29283) matches that of the mandible. The incisor-canine occlusion was locked. It is typical of a semi-erect axial skeleton and does not correspond to the erect occlusal-postural equilibrium of a hominid, either *Australopithecus* or *Homo*. If a new species can be defined, there is no reason to decree that it defines the genus *Australopithecus*, which would have the sole aim of ticking the hypothetical boxes of a gradualist doctrine, whereas the evolutionary modalities of the regulation of embryonic architectural dynamics are saltationist.

Facial fossils attributed to *anamensis* have nothing to convince of Hominian differences with *Ardipithecus*; they rather attest to the existence of a species of *Ardipithecus* in a humid and wooded environment, living a thousand kilometers away from the *ramidus* species. Mandible fragments were also collected in Fejej (southern Ethiopia) (de Lumley 2007).

> These discoveries also made it possible *to question the classic Anglo-American tendency to attribute to Hominids, or even more precisely to Australopithecus, all hominoid pieces dated between 9 and 4 Ma*. We realize today that the diversity is much greater than we thought, hence the discussions; we can, however, try to sketch a slightly more complete scenario today. So, who are these oldest Rift Hominids? (Senut, *L'évolution de l'homme*, 2009, author's translation)

This critical look confirms the thinking of the school of the *Muséum national d'Histoire naturelle*, forged by five generations with experience in Africas and hominids, with Pierre Teilhard de Chardin, Camille Arambourg, Yves Coppens and Brigitte Senut, inventor of several fossils, including *Orrorin tugenensis*. The French paleontologist, a specialist in the appendicular skeleton of Hominids and *Australopithecus* in particular, is well aware of the taxonomic revisions of fossils classified as Hominin, which are families of great apes.

6.2.4. *Australopithecus and Homo contemporaries at 4 million years?*

Pierre Teilhard de Chardin did not follow Robert Broom in his linear vision of the *Australopithecus* ancestors of the genus *Homo*. This lineage was only a hypothesis, all the less supported as fossils of the appendicular skeleton confirmed a human lineage already present at 4 million years. As Brigitte Senut reminds us, the humerus of Kanapoi "is really very human and differs from *Australopithecus*", as does the tibia, whose knee and ankle extremities correspond to the posture of *Homo sapiens* today, and not that of an *Australopithecus*:

> The proximal condyles, which articulate with the distal end of the femur, are concaved like the condyles seen in anatomically modern *Homo sapiens*. Additionally, similar to modern humans, the proximal tibia is wider near the knee joint to allow the deposit of extra spongy bone, which acts as a shock absorber during bipedal locomotion. This extra deposit of spongy bone can also be seen in the medial malleolus, which articulates with the ankle. Overall, the partial tibia is more human-like than it is ape-like. The fact that the shock-absorbing mechanisms so closely mirror the same mechanisms seen in modern humans led paleoanthropologists to believe that this species was a habitual biped". (Hominid Fossil Repository, *Australopithecus anamensis*, online)

The paleontologists who collected it nevertheless decided to make it an *Australopithecus*, whereas the French paleontologist, an expert in this field, was categorical:

> A very human, almost complete tibia, indicates a well-established bipedality in this Hominidae, different from that of the classical Australopithecus (…). The presence of a locomotor skeleton oriented towards bipedalism allows us to consider this material as a 4 Ma old testimony of the human lineage. Thus (the species of Kanapoi) could belong to the *Praeanthropus* genus rather than to the genus *Australopithecus* and would testify to the ancient presence of the human lineage. (ibid., author's translation)

The tibia corresponds to a human-type bipedality rather than an Australopithecine one, but what is important above all is what the joint reveals about the straightening of the endoskeleton, because the distribution of stresses according to verticality corresponds to an embryonic straightening different from Australopithecine. The regulation of embryogenesis has reshaped "great ape" morphogenesis according to the cranio-spinal gradient of the embryo. The dorsal cord became vertical, so the contraction of the basi-sphenoid was necessarily more pronounced than that of the great ape, but it was not that of an *Australopithecus*, whose cranio-spinal skeletal anatomy is completely distinct.

As in Ethiopia, care must be taken not to create a half-monkey/half-human chimera by applying the gradualist linear doctrine without regression. If the humerus and the tibia of Kanapoi are those of the genus *Homo*, this emergence does not pose any problem phylogenetically, because the threshold of complexification of the neural tube and its effects on cranio-sacral regulation can generate both embryogenesis, or even more.

Allia Bey's fossils are essentially isolated teeth, plus a radius fragment, a half mandible with the molars (KNM-ER 30745), as well as an incomplete half mandible with the canine root and the two premolars. The maxillary first premolar is bicuspid and molariform, while an isolated lower canine (KNM-ER 30731) shows a gracile root and a low crown with apical wear typical of an end-to-end occlusion (Ward *et al.* 2001). Unlike the Kanapoi faces, these 3.9 Ma fossils are hominin, which is consistent with the *Australopithecus* base of Aramis of the same geological age.

The break with the dogma of the great ape-*Australopithecus*-*Homo* linearity became more pronounced following the collection of an incomplete skeleton in Ethiopia, dated 3.58 Ma, nicknamed "Kadanuumuu" (KSD-VP-1/1). The study has been published in detail (Haile-Selassie *et al.* 2010). The individual was about 1.60 m tall. The appendicular skeleton and vertebrae have nothing in common with the bones of *Australopithecus* dated 3.2 Ma, collected 50 km away and classified as *afarensis*. The legs are longer, the locomotion on the ground was a permanent bipedality by stride, close to Man and not swayed by small steps like the *Australopithecus* Lucy (Afar Locality 288). In spite of the facts, the collectors decided to give it the name of *Australopithecus*, which provoked reactions, readable at the time on the site "Hominins":

> The discovery of Kadanuumuu shows that we are dealing with another hominin (more or less exclusive biped, longer lower limbs, trunk and scapula different from *Au. afarensis*, himself a biped and climber). This proposal of the existence in the Pliocene, around 3 to 4 Ma, of two lines of hominines, one with a more human skeleton and bipedal over long distances (which would be represented by Kadanuumuu), and the other also bipedal, but in a different way, with a smaller skeleton and a good climber (represented by Lucy) is that of Brigitte Senut, professor at the National Museum of Natural History, Paris, among others.

Kadanuumuu's skeletal morphological organization plan has nothing in common with *Australopithecus*, and if he is already human, the nomenclature must call him such.

The scenario was repeated with a 3.2 Ma fourth metatarsal discovered in the Afar (AL 333-160), arbitrarily attributed to *Australopithecus afarensis*, because anthropologists note a curvature and length close to the arch of modern Man (Ward *et al.* 2011), whereas *Australopithecus* had no arch. Yvette Deloison (2004) is equally critical of the attribution to this genus in the name of a gradualist dogmatism that does not conceive of *Homo* anatomy over 3 Ma old.

Fergusson pointed out that among the fossils of Hadar Hominids, some were human:

> The assignment of fossil hominoid jaws from the Plio/Pleistocene of Hadar to a single genus, *Australopithecus* Dart, 1925, is a misnomer. They are morphologically unrestricted to and inconsistent with the diagnosis and evolutionary trend of *Australopithecus*. The morphological pattern of four large jaws is indeed australopithecine and similar to *A. africanus* Dart, 1925, but six small jaws reveal a pre-habilis stage of dental development early in the *Homo* lineage. (Fergusson, *Revision of fossil hominid jaws from the Plio/Pleistocene of Hadar*, 1984)

The differences are clearly at the level of the morphogenetic plan which is structured from embryogenesis. *Homo* and *Australopithecus* are two embryogenesis verticalized according to divergent homeotic rhythms and regulations; giving these fossil bones the name *Australopithecus* in the name of a dogma, instead of starting by understanding the modalities of their development and recognizing distinct patterns, is to expose oneself to embryological nonsense with serious consequences for the understanding of hominization processes. To postulate the filiation between *Australopithecus* and *Homo* before knowing these modalities from the current species brings an additional difficulty and leads one to look in the wrong direction, in seeking an ecological engine justifying the transformation of the swaying walk of *Australopithecus afarensis* into a step-by-step walk. *Since permanent bipedalism is neither a cause nor an end of these transformations, the real evolutionary problem goes unnoticed.* Walking in strides makes it possible to cover longer distances if required, but it is not clear how it would be incompatible with living in the environment, which was the original context (wetlands which remained densely forested thanks to groundwater), since the anatomical transformation is the consequence of the complexification of the homeotic gene's regulation.

The need for a prefix *prae* to *anthropus* is a relic of the gradualist doctrine. The emergence of *Homo* embryogenesis (which should not be confused with *Homo sapiens* or Sapiens embryologically speaking) before 3 Ma from great ape gametogenesis is possible and does not require *Australopithecus* embryogenesis as they are so close. Taxonomic disagreements may seem derisory, but to bring all the fossils together in a single organization plan would be a mistake at the expense of our species, here and now, preoccupied by the disorders linked to embryonic axial straightening and incontestable phylogenetic determinism transmitted by gametes. The taxonomy of fossils deserves more reflection when it comes to the processes of emergence of the genus *Homo* and thus of *Sapiens*, living today.

6.3. Axial straightening, cranio-palatal balance and occlusion of Hominids *sensu stricto*

6.3.1. *The Pliocene and Lower Pleistocene African Hominids*

6.3.1.1. *The mandible*

The fossils of mandibles, particularly numerous, are mostly fragments of mandibular bodies, sometimes indeterminable between the genera *Australopithecus*, *Paranthropus* and *Homo*. The origin of the fossils is detailed in the grid below.

Ethiopia	Afar (Afar Locality, AL), Omo
Tanzania	Laetoli (LH), Olduvai (OH), Peninj
Kenya	Est Rudolf (KNM-ER)
South Africa	Makapansgat (MLD), Swartkrans (SK), Taung, Kromdrai (TM)

The collection consisted of 35 mandibles, six of which were non-adult, sufficiently well preserved to take measurements, sometimes symmetrically (Table 6.2). Comparison with the great apes and *Homo sapiens* made it possible to distinguish *Australopithecus* from *Homo* and *Paranthropus* from *Australopithecus*.

Phase	n	Specimen
Phase 0 Neonatal stage no teeth	0	–
Phase 1 Temporary teeth	1	AL 333-43
Phase 2 Eruption from M1 to M1 in place	3	LH 2, Taung, KNM-ER 820
Phase 3 Eruption from M2 to M2 in place	2	MLD 2 and AL 128
Phase 4 M3 in place, adult	29	

Table 6.2. *Number of fossils and specimen reference by dental phase*

The measures are as follows:

Height of the symphysis (l), basilar length from the pre-angular notch to the gnathion (L), the symphysis angle (α) and the basilar angular deviation (β).

The measures are presented in Table 6.3.

Specimen	Taxon	Age Ma	l	L	D	α	β
LH 4 holotype	*afarensis*	3.6	35	60	69	70	105
LH 2*	*afarensis*	3.6	30	43	57	82	–
MLD 40	*africanus*	3.33	38	69	–	–	120
MLD 2*	*africanus*	3.33	33	60	68	69	118
AL 400	*afarensis*	3.2–3	39	70	78	68	–
AL 198	*afarensis*	3.2–3	37	61	70	70	108
AL 207	*afarensis*	3.2–3	32	67	74	67	107
AL 288 Lucy	*afarensis*	3.2–3	31	60	67	68	103
AL 266	*afarensis*	3.2–3	30	63	72	70	105
AL 128*	*afarensis*	3.2–3	–	57	70	76	105
AL 333-43*	*afarensis*	3.2–3	22	37	50	85	107
AL 277	*afarensis*	3.2–3	42	–	–	–	–

AL 333 w 60	*afarensis*	3.2–3	50	67	72	65	105
Omo 860	*afarensis*	3.36–2.85	40	76	82	65	102
Omo 18	*afarensis?*	2.7–2.3	38	70	80	70	110
Omo 57	*afarensis*	2.3	43	75	83	67	110
Omo L 7 A	*afarensis*	2.27–2.19	56	75	82	66	115
OH 13	*Homo habilis*	1.7	25	66	–	–	109
OH 22	*Homo*	1.25–0.78	34	65	82	77	109
KNM-ER 1802	*Australo* vs. *Homo*	2 –1.94	39	65	70	65	109
KNM-ER 1482	*Australo* vs. *Homo*	2–1.94	38	60	70	71	112
KNM-ER 1805	*Australo* vs. *Homo*	1.74	32	70	77	67	–
KNM-ER 820*	*Homo*	1.7–0.8	30	51	70	84	107
KNM-ER 730	*Homo*	1.7–0.8	30	65	80	76	99
KNM-ER 810	*A. robustus*	1.77	47	72	80	67	108
KNM-ER 729	*A. robustus*	1.55–1.49	50	84	88	72	105
KNM-ER 1468	*A. robustus*	1.55–1.49	50	70	–	–	–
KNM-ER 801	*A. robustus*	1.59	–	80	–	–	–
KNM-ER 725	*A. robustus*	1.55–1.49	34	87	–	–	–
KNM-ER 992	*Homo ergaster*	1.5	37	73	90	77	110
KNM-ER 404	*A. robustus*	1.42	40	70	–	–	–
Taung *	*A. africanus*	3–2.6	25	38	45	73	118
SK 23	*Paranthropus*	1.8–1.5	50	65	–	–	96
SK 15	*Homo*	1.5	30	69	89	80	97
Peninj	*Paranthropus*	1.56–1.70	50	73	94	80	96

Table 6.3. *Linear and angular values of the mandibles of Australopithecus afarensis, africanus, Paranthropus and Homo no sapiens. Taxa with * are non-adult stages. Dating is based on the works by Cerling* et al. *(2011) and Wood and Constantino (2007)*

The shape of the mandible changed dramatically in the infantile stage, with a widened alveolar arch, pronounced incisor-canine frontalization, the appearance of the symphyseal void and a symphyseal angle much more closed than a great ape. The symphysis is more upright (Figure 6.9).

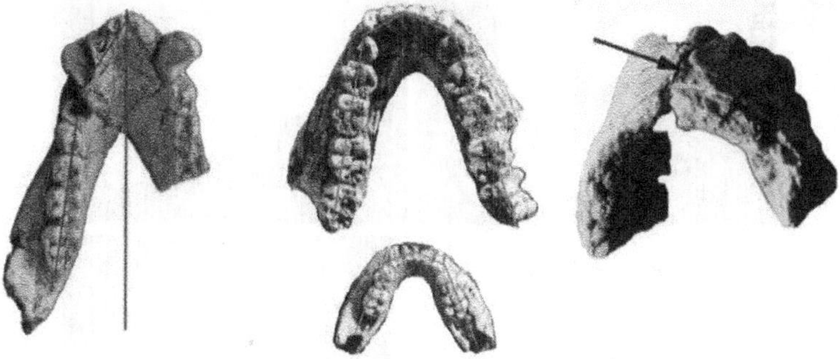

Figure 6.9. *Mandibles in occlusal view of* Dryopithecus africanus, *(left) and* Australopithecus afarensis, *young and adult, with the frontalization of the incisor-canine alveoli, the canine at occlusal height of the other teeth and the first molariform premolar. The arrow indicates the location of the ossified symphyseal void on a mandible of* Australopithecus afarensis *(photos: A. Dambricourt Malassé, different scales). For a color version of this figure, see www.iste.co.uk/dambricourt/embryogeny.zip*

The first deciduous molar (dml) forms five cusps and not two, it is not canine-like (Figure 6.10); the crown of the canine is not higher than the molars. The lower canine is still in front of the upper, but the dental articulation has completely changed; the canine lock no longer exists (Figure 6.11).

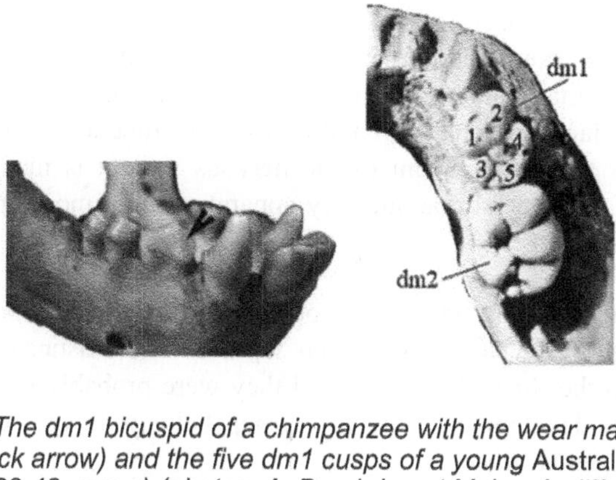

Figure 6.10. *The dm1 bicuspid of a chimpanzee with the wear mark of the upper canine (black arrow) and the five dm1 cusps of a young* Australopithecus *of Afar (AL 333-43, arrow) (photos: A. Dambricourt Malassé, different scales)*

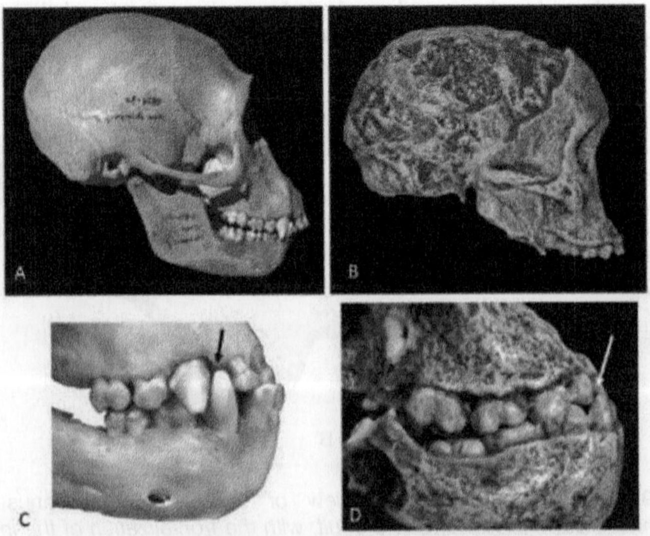

Figure 6.11. *Comparison between a chimpanzee (A and C) and an* Australopithecus *in mixed dentition (child of Taung, M1 in place, B and D), the incisor-canine intercuspidation in hominins has no canine lock (photos: A. Dambricourt Malassé, different scales)*

The adult basilar angular deviation shows two groups: one is composed of individuals whose angle is as closed as in the great apes, while the infantile stage is as open as in the latter and as seen in *Homo sapiens*. The angular trajectory thus indicates an anteroposterior growth, that of the alveolar-dental arch, faster than the transverse growth. Cerebro-cerebellar growth which corresponds to psychomotor development slows down earlier in the great ape than in humans, with the development of masticatory functions. These modalities were found in *Australopithecus*, facial prognathism and masticatory functions developed, whereas in *Homo sapiens* the development of the nervous system is maintained with the disappearance of a powerful manducatory apparatus and a more open adult basilar angular deviation.

These characteristics are observed in other adult mandibular bodies with a more open basilar angular deviation like *Homo sapiens*. The masticatory functions are more gracile than the *Australopithecus* and they were probably delayed in favor of cerebro-cerebellar development. The first group belongs to the Australopithecine, the second to the genus *Homo*.

Mandibles from Rudolf's East are classified in the hominid group but without further details. The basilar angular deviation makes it possible to slice, for example KNM-ER 1802, 1482 and 1805 are *Australopithecus*. Among the *Australopithecus* of Rudolf's East, one group is composed of large individuals (long mandible) with strong sexual dimorphism (height of the symphysis). The nomenclators named it *Paranthropus boisei*. In reality, this genus, previously named *Zinjanthropus*, had a pronounced embryonic contraction, but was different from *Australopithecus* and Man, as we will see, with a strong frontalization of the petrous pyramids and a widening of the base that does not include these too narrow mandibles. These mandibles were rather those of a large East African *Australopithecus* species, *Australopithecus aethiopicus* or *Australopithecus robustus* (Dambricourt Malassé 1987).

The mandibular frontalization corresponds to that of the maxilla (Figure 6.12).

Figure 6.12. *On the left, the palate of a Miocene great ape (*Dryopithecus*) and on the right that of a* Hominidae Australopithecus aethiopicus *(KNM WT 17000). The alveolar arch is frontalized and the canine alveolus is small (photo: A. Dambricourt Malassé). For a color version of this figure, see www.iste.co.uk/dambricourt/embryogeny.zip*

We have seen (section 5.2.3) that the integration of the alveolar arch (pars alveolaris) into the transverse growth of the pars basilaris corresponds to an anterior widening of the floor of the middle cerebral fossa, in other words to a lateral encephalization of the temporal lobes. This is effectively verified with the endocranial cast of the Taung child described below.

6.3.1.2. *The gain in axial verticality and craniofacial contraction of Australopithecus*

Collections of complete Australopithecine skulls with their mandibles have been enriched over about 20 years, but the straightening of the clivus, the pyramids and their frontalization are still not the unifying evolutionary problem. The lack of interest in the first trajectories of morphogenesis can be explained, among other things, by the scarcity of complete and slightly deformed skulls at the end of the 1980s. These collections were limited to *Australopithecus africanus* (Taung and Sterkfontein Sts 5) and *Australopithecus aethiopicus* KNM WT 17000 (Kenya). These more recent discoveries concern six skulls, two of which were found in South Africa, with Drimolen's *Paranthropus robustus* (DNH 7) and Sterkfontein's *Australopithecus prometheus* (Little Foot Stw 573), as well as four skulls of *Australopithecus afarensis* discovered in Ethiopia: two adult skulls deformed with A. L. 444-2 and A.L. 822-1 dated 3.2 Ma (Kimbel and Rak 2010), the skull of Woranso-Mille dated 3.8 Ma (MRD VP-1/1) and the child of Dikika dated 3.3 Ma. The latter specimen is exceptional because the upper part of the skeleton with the cervical spine and a scapula is preserved. The skull is still in articulation with the atlas, while the scapula shows an arboreal gorilla-like conformation, corroborated by the phalanxes. These six skulls show the absence of a canine lock.

The study of craniofacial contraction in *Australopithecus* was made possible thanks to the availability of CT scans by Phillip Tobias. The protocols could be applied to Taung and two Sterkfontein skulls, Sts 5 and Stw 505, of Member 4 of the sedimentary filling dated between 2.6 and 2 Ma (Herries *et al.* 2013).

Below is a reminder of the values of paninae and *Homo sapiens* (Table 6.4).

Taxon	sph	σ	rsph	σ	A1	σ	A2	σ	a1	σ	a2	σ	a3	σ
Sapiens	127	5	119	5	49	8	70	8	83	9	115	9	48	5
Pan	155	6	132	4	88	8	105	8	144	5	162	4	80	2
Gorilla	157	10	137	8	–	–	–	–	–	–	–	–	–	–

Table 6.4. *Mean values and standard deviations of sphenoidal angle (sph), sphenoidal rotation (rsph), internal (A1 and A2) and external (a1, a2 and a3) contractions*

6.3.1.3. *Taung's child*

Taung did not retain the cranial base, but the endocranial cast and the assembly of the scans allowed the osseous equivalences of basion and sphenoidal planum to be located, with a good approximation of the position of the ephippion (Figures 6.13 and 6.14). The biometrical protocol could be applied (Table 6.5).

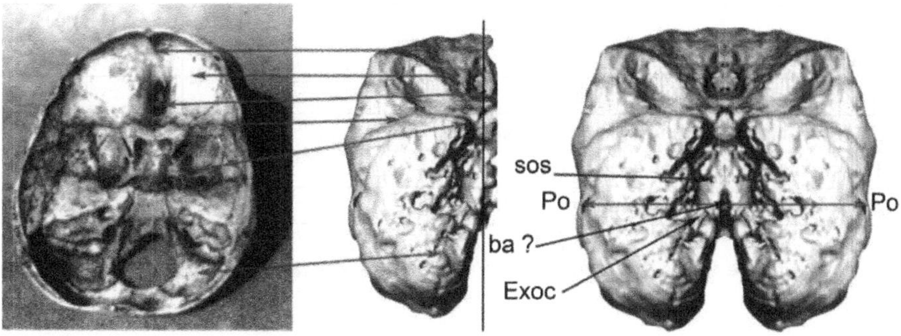

Figure 6.13. *Reconstruction of the Taung endocranial cast in lower view by symmetry and interpretation of bone equivalences thanks to a comparison with the young gorilla (see Figure 5.7): Ba: basion, Exoc : exoccipital, Po: porion, SOS: spheno-occipital synchondrosis. The basion is very close to the straight line which joins the two porions and further forward than in great ape. We recognize the body of the sphenoid with a triangular shape and the basi-occipital. An imprint of the exoccipital seems preserved, useful for locating the foramen magnum (Dambricourt Malassé 2011b). For a color version of this figure, see www.iste.co.uk/dambricourt/ embryogeny.zip*

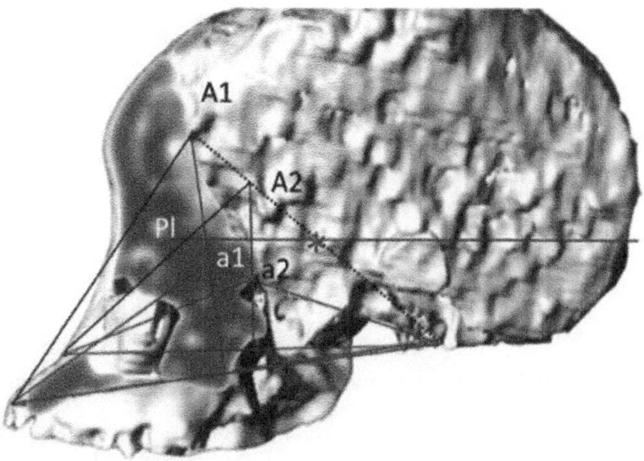

Figure 6.14. *Application of the contraction angle measurement protocol. The position of the ephippion (asterisk) for the sphenoidal angle is strongly constrained by the basi-cranio-facial structural environment. Pl: sphenoidal planum (Dambricourt Malassé 2011b). For a color version of this figure, see www.iste.co.uk/dambricourt/ embryogeny.zip*

Angle	Degrees
Sphenoidal	140
A1: Internal contraction/prosthion	84
A2: Internal contraction/anterior nasal spine	100
a1: External contraction/prosthion	128
a2: External contraction/anterior nasal spine	142

Table 6.5. *Sphenoidal and cranio-facial angular values of Taung*

The integration of the mandibular alveolar arch in the lateral growth trajectory of its basilar part implies a widening of the palate due to the lateral expansion of the anterior end of the temporal lobe. Comparison of a chimpanzee's X-rays of the same dental stage, with the lower view of Taung's endocranial cast, symmetrically reconstructed, confirms a fronto-lateral expansion of the anterior end of the temporal lobe (Figure 6.15).

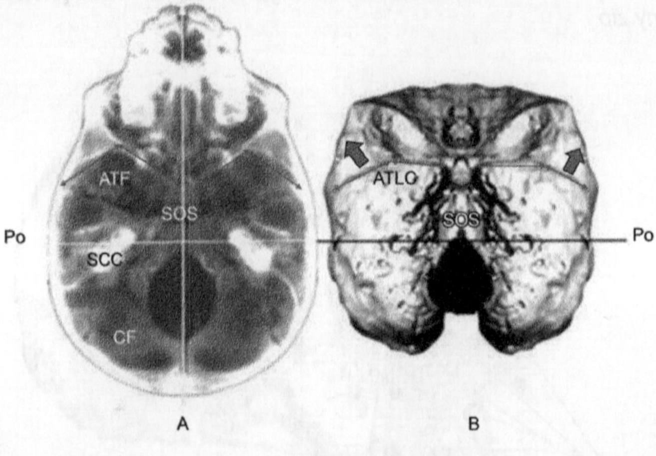

Figure 6.15. *Comparison of the anterior end of the temporal lobe between a chimpanzee (A) and Taung (B) at the same dental stage. SCC: semi-circular canals; ATLC: anterior temporal lobe cast; ATF: anterior temporal fossa; CF: cerebellar fossa; SOS: spheno-occipital synchondrosis; Po: porion. The arrows show the fronto-lateral expansion of the anterior end of the temporal lobes in the young* Australopithecus *(photos: A. Dambricourt Malassé). For a color version of this figure, see www.iste.co.uk/dambricourt/embryogeny.zip*

6.3.1.4. *Sterkfontein 5 (Sts 5)*

The pre-sphenoid is slightly raised, while the well-preserved post-sphenoid shows the very clear gain in verticality, incomparably more contracted than a great ape. However, unlike *Homo sapiens*, the basi-occipital is not aligned with its straightening. Consequently, the pharyngeal or exocranial face of the clivus is much less vertical (Figure 6.16).

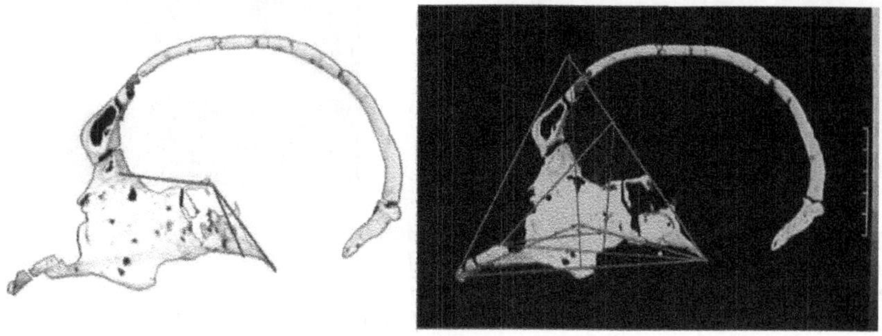

Figure 6.16. *CT scan of* Australopithecus africanus *Sts 5 – courtesy of Phillip Tobias in 2003 – on the left with the sphenoidal rotation angle and the sphenoidal angle, and the contraction angles on the right (A. Dambricourt Malassé 2011b). For a color version of this figure, see www.iste.co.uk/dambricourt/embryogeny.zip*

Sterkfontein's sphenoidal angle and sphenoidal rotation angle are more open than *Homo sapiens:* the sphenoidal angle of *Australopithecus afarensis* A.L. 822-1 is 142°. The internal and external contraction angles of Sterkfontein 5 are much greater than *Homo sapiens* due to the prognathic development of the face (Table 6.6).

Angle	Degrees
Sphenoidal	147
Sphenoidal rotation	133
A1: Internal contraction/prosthion	62
A2: Internal contraction/anterior nasal spine	79
a1: External contraction/prosthion	142
a2: External contraction/anterior nasal spine	158

Table 6.6. *Sphenoidal and cranio-facial angular values of Sterkfontein 5*

A geometrical trace in the median sagittal plane makes it possible to visualize the gain in embryonic verticality of the dorsal cord and thus of the supra-chordal neural tube (Figure 6.17).

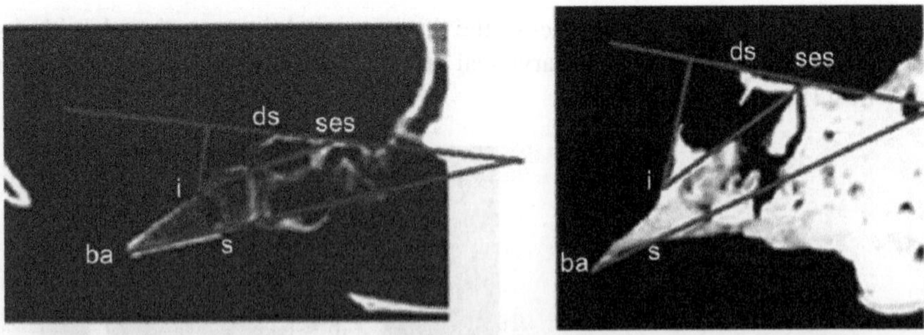

Figure 6.17. *Left: Pan skull base, right:* Australopithecus africanus *Sts 5. Geometric visualization of the embryonic contraction of the basi-sphenoid showing the gain in verticality of the axial endoskeleton. ba: basion, ds: anterior edge of the sella turcica (dorsum sellae), i: endocranial face of the spheno-occipital synchondrosis, s: exocranial face of the spheno-occipital synchondrosis, SES: location of the spheno-ethmoidal synchondrosis (Dambricourt Malassé 2004, 2005). For a color version of this figure, see www.iste.co.uk/dambricourt/embryogeny.zip*

The gain in verticality of the axial endoskeleton is spectacular. The angle of rotation of the basi-sphenoid is undisputable proof of a change in the embryonic neural dynamics of the great ape. The genetic regulation of the neural crest is prolonged, as revealed by the formation of the five cusps of the first deciduous molar. The embryonic chronology is very precise. The evolution of the regulation of homeotic genes has evolved in the same direction as in humans, with an extension, and, therefore, a heterochrony, of neurogenesis. The angle of sphenoidal rotation means a closure of the IV ventricle to the effects on sphenoidal synchondroses, which are longer than in the great ape, but which cease earlier than in humans. Second, the tensions in the spheno-occipital synchondrosis are different since the basi-occipital has not aligned with the slope of the basi-sphenoid. These early discrepancies increased with development; *Australopithecus* maintained facial hormonal growth inherited from the great ape. This difference in the evolution of the regulation of the homeotic genes is confirmed by the semi-circular channels that retain a simian conformation.

These extremely precise details on the first weeks of embryogenesis masterfully demonstrate that the appearance of a Hominid is not finalized by locomotion, bipedality and verticality. If the straightening of the basi-sphenoid was "finalized" by verticality, and therefore by the threshold of exclusive bipedal locomotion, the semi-circular canals would be the same in *Australopithecus* and *Homo*. However, we have shown that the canals are human while the rotation of the basi-sphenoid is not effective. The morphogenesis of the semi-circular canals is done by apoptosis (cell death) under the control of homeotic genes.

The emergence of Hominids thus corresponds to different processes of complexification of neurogenesis, with specific cranio-spinal homeotic regulations resulting in the neo-formation of neural networks between the cerebellum and the brain (Dambricourt Malassé and Lallouet 2009).

Human bipedalism is no more "advanced" than that of an Australopithecus *or a* Paranthropus, *since this evolution is not that of bipedalism. Neither is "more advanced" than the other. They are nervous systems with different degrees of verticalization from the brain stem. They are therefore carried and moved by the two lower limbs according to occluso-postural control modalities specific to each morphodynamic system.* Homo sapiens *is more advanced in its endoskeletal verticality and its bipedal equilibrium is therefore different. This is not a bipedalism which has evolved, but rather the embryonic verticalization and the complexification of the neurogenesis in particular have been amplified.*

6.3.1.5. *Vectors of occipital squamous growth*

For the first time in the evolution of embryonic straightening, the occipital squamous follows the intra-uterine vectors by continuing the coiling (Figure 6.18). The cerebellum necessarily develops longer with space forward and downward, and due to the control of postural balance of the verticalized and therefore more unstable endoskeleton. *The result is an obligation for the central nervous system to restructure itself to complexify the connections between the brain and the cerebellum, with the need for the latter to integrate itself into the analysis loop.*

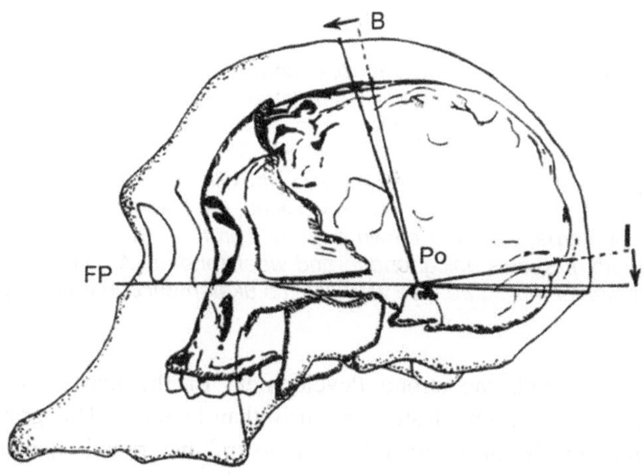

Figure 6.18. *Growth vectors of vault and occiput identified in the Frankfurt Plane centered on the porion. Sts5 and Taung (Dambricourt Malassé 1987, 1993)*

Since the skull is no longer cantilevered as in the semi-erect great ape, the occlusion is modified with the disappearance of the incisor-canine lock (Figure 6.19). The new occlusal-postural equilibrium is freed from the horizontal tensions generated by the semi-erect orientation of the axial endoskeleton.

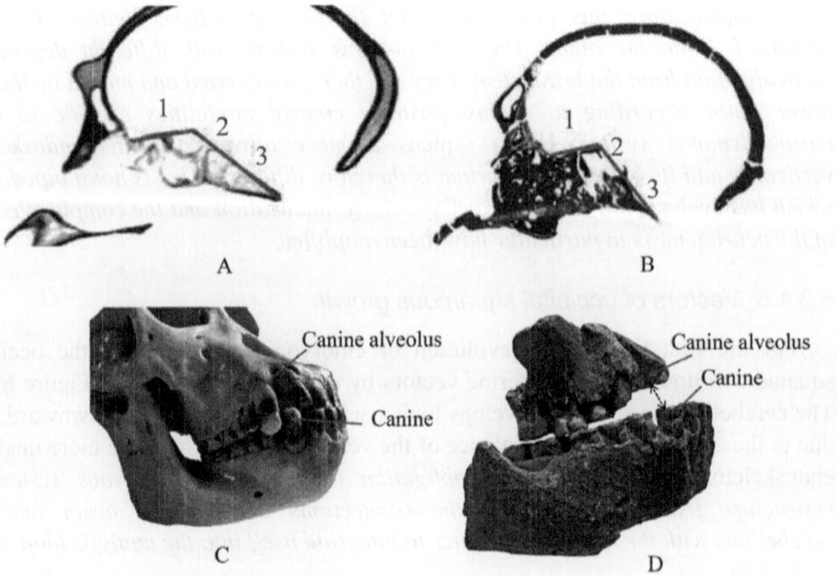

Figure 6.19. *Comparison between basi-sphenoid rotations and incisor-canine occlusion of a chimpanzee and an adult* Australopithecus. *The gain in verticality is visualized by the staggering of the three fossa (1, 2, 3). A: chimpanzee, B:* Australopithecus africanus *from Sterkfontein Sts 5, C: the incisor-canine lock in adult chimpanzee, D:* Australopithecus prometheus *(maxilla MLD 9 and mandible MLD 18) from Makapansgat, the occlusion shows no locking, the lower canine remains in front of its maxillary counterpart, but the crown does not protrude beyond the occlusal plane of the other teeth. It shows apical end-to-end wear (photos: A. Dambricourt Malassé). For a color version of this figure, see www.iste.co.uk/dambricourt/embryogeny.zip*

On the other hand, the strong development of the masticatory muscles is preserved and the face grows faster anteriorly than laterally. This growth inherited from great apes results in a forward trajectory of the frontal squamous with the formation of the supraorbital torus; it dissociates the external table from the internal table which remains in contact with the frontal lobes. This trajectory, inverse to the occipital squamous, corroborates the rapid closure of the basilar angular deviation of the mandible. It gives the adult mandibular arch a more flared shape than that of a great ape, but narrower than in the juvenile stage.

The heterochrony of development does indeed concern a prolongation of neurogenesis as soon as the neural tube is formed. Further proof is given by the delay of hormonal growth of the face with respect to the dentition; in fact, Taung is already in mixed dentition (first molar M1 in place), while the supraorbital torus is not yet visible. The psychomotor development of an Australopithecine is thus in no way comparable to that of a species of great ape. Everything indicates a longer duration of neuronal development and a learning of the psychomotor control of balance which is necessarily more complex, with the cerebellum being forced to integrate into the loop. *A reorganization and complexification of the neural network between neo-cortical areas of the brain and cerebellum is a first-order inference.* A comparison of volume ratios between casts of the Taung brain and cerebellum with the chimpanzee could test for cerebellar neo-corticalization.

6.3.1.6. *Cranio-palatal balance: the postural breakthrough with the great apes*

CT scans and exocranial photography at natural scale have made it possible to trace the first analysis of the cranio-palatal balance of an *Australopithecus* with Sterkfontein Sts5 and Sterkfontein Stw 505 (Figure 6.20). The comparison with Figure 5.9 shows the conservation of balance in the adult stage, consistent with the conservation of sphenoidal rotation.

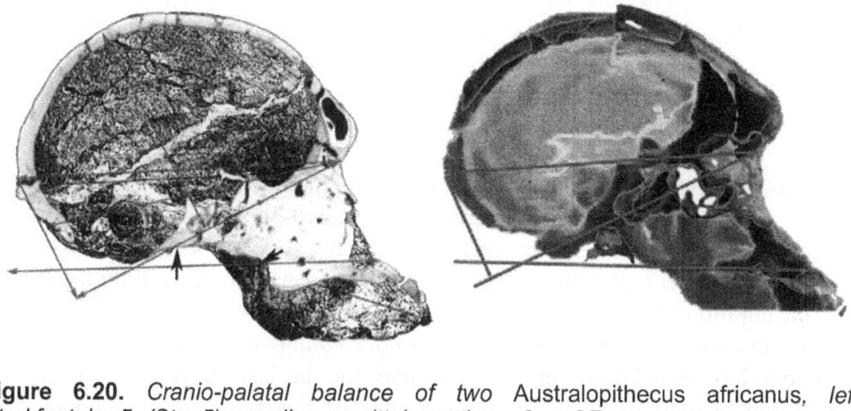

Figure 6.20. *Cranio-palatal balance of two* Australopithecus africanus, *left Sterkfontein 5 (Sts 5), median sagittal section of a CT scan projected on the exocranial profile. Original scanner, original construction. On the right Sterkfontein 505 (Stw 505), original virtual reconstruction by superposition of a scanned section and a CT scan (Dambricourt Malassé 2011b). For a color version of this figure, see www.iste.co.uk/dambricourt/embryogeny.zip*

The straightened symphysis suggested that the cranio-palatal balance was maintained by a verticalization of the axial endoskeleton and thus a more contracted basi-sphenoid.

Permanent bipedality is a morphological criterion relevant to distinguishing a Hominid from a great ape, but it is not the cause of the distinction. This criterion is the threshold of endoskeletal verticality and numerous morphodynamic and architectonic changes along the cranio-caudal embryonic axis. The cranio-palatal balance is the most obvious criterion of a change in cranio-caudal embryonic dynamics, it allows us to define a true evolutionary distinction between the Hominid and the architectural biodynamics that preceded it. This limit does not offer any possible confusion: either the cranio-palatal balance disappears, because the flexural dynamics cease, or it is preserved, which is only possible as a result of a remodeling of cranio-palatal architectural constraints since embryogenesis. It imposes a permanent bipedal equilibrium on the ground, from childhood to adulthood, which has nothing in common with the occasional infantile stage of the great apes.

The great innovation is not bipedalism, present for millions of years in the lineage of the gibbons. It is the irreversible disappearance of locomotion on the ground by leaning on the anterior extremities, it is the threshold of the disappearance of the quadruped locomotor balance in primates, which gives primacy to the cerebellum in a vertical endoskeletal posture and required rapidly learning to control the movements of the head and neck. Telencephalization would not be observable, as the study of the Dikika endocranial cast concludes[1]. However, Taung and the mandibles of *Australopithecus afarensis* show a fronto-lateral expansion of the anterior end of the temporal lobe. The innovation corresponds to a complexification of neurogenesis, which has the effect of a more pronounced closure of the VI ventricle, and an inevitable complexification of the neural networks between the cerebellum and the brain that is not visible on endocranial casts.

Thus, the Hominid is the first threshold of straightening of the supra-chordal neural tube to have reached verticality, generating an acceleration of the complexification of the cerebrocerebellar neural networks, as well as the exceeding of a "cognitive ceiling" limiting the amount of information received by the sensors and the duration of the analyses. *It is the instability of the cerebellum that generates a jump in complexity in the reflection of information.* Its integration in cognitive functions has been known in *Homo sapiens* for the last 30 years (Schmahmann 1991) and is most certainly the consequence of this reorganization imposed by verticality, itself a consequence of phylogenetic evolution of embryonic complexity. The cephalic pole is thus reorganized like the entire axial and appendicular skeleton, without altering the post-natal arboreal functions, which do not depend on the organization of the embryo. *Australopithecus* and *Homo* could thus preserve these post-natal adaptations.

1 The endocranial imprint of the Dikika child reveals an organization of the brain similar to that of a monkey and no characteristics derived from that of humans. (Philipp Gunz, MPI EVA, Leipzig).

This break with the quadrupedy does not lend itself to gradualism, variability within the limits of the same embryonic organization or statistical measurements of Gauss curves; it is saltationist and macro-evolutive, linked to a reorganization of morphogenesis, a complexification of the nervous system and an elongation of the neurulation phase which ends with the definitive acquisition of axial verticalization of the endoskeleton.

6.3.1.7. *Frontalization of the petrous pyramids*

Dean and Wood (1982) showed the exocranial frontalization of petrous pyramids in *Australopithecus* and *Homo*, without correlating it to the rotation of the sphenoid. Our statistics established between *Homo sapiens* and *paninae* show a strong correlation between internal craniofacial contraction, rotation of the basi-sphenoid, and frontalization of the petrous pyramids. In other words, frontalized pyramids correspond to a gain in axial verticality, or to a sphenoidal rotation angle more closed than that of a *Pan*, in relation to up and down rotation dynamics on their major axis.

Australopithecus aethiopicus: KNM-WT 17000 (2.52 Ma)

The species name *aethiopicus* was created by Camille Arambourg and Yves Coppens in 1967 for a mandible from the Omo Valley (southwestern Ethiopia), Omo 18, discovered in 1967 in levels dated to about 2.52 million years ago. The genus created for this mandible was *Paraustralopithecus*. The angular deviation is 70°, like AL 266, smaller in length and lower in height (the symphysis is shorter), the symphysis angle of 110° is more open than AL 266 (105°) and therefore more prognathic. It is a mandible of *Australopithecus*.

The fossil KNM-WT 17000, nicknamed “Black Skull” (see Figures 6.12 and 6.21), is among the most complete and best preserved specimens in East Africa (west Turkana). The good quality cast shows that there is no anatomical contact point between the face and the neurocranium. However, this solution of continuity has induced a positioning error which has the effect of raising the face. Since the fossil was not scanned, the endocranial base and the sphenoidal angle are not known.

The pyramids are frontalized and show a difference of prime importance for the stability of the verticalized area of the base. Their anterior end is in contact with the spheno-occipital synchondrosis (SOS), whereas that of the great apes is forward (Figure 6.21).

Figure 6.21. *Comparison in exocranial view of the orientation of the pyramids of KNM-WT 17000 and an adult gorilla. The major axis of the pyramids is clearly frontalized; they maintain the SOS on either side of the plane of symmetry (photo: A. Dambricourt Malassé). For a color version of this figure, see www.iste.co.uk/dambricourt/embryogeny.zip*

The pyramids are integrated into a cranio-sacral dynamic system dominated by verticality and forward and downward rotation vectors, with the SOS as its seat.

This fossil has been classified in the genus *Paranthropus* because of secondary sexual characteristics, such as large size and significant development of the crests on the top of the skull. They are formed by the aponeuroses of the temporal muscles that support the mandible; it is a formation related to mastication independent of the flexion of the base. However, the base of the endocranium has not been studied, but its exocranial – or pharyngeal – face is totally different from *Paranthropus* SK 48 and *Zinjanthropus*, renamed *Paranthropus boisei*. This skull is an *Australopithecus robustus* and the closure of the IV ventricle had nothing in common with *Paranthropus*.

6.3.1.8. *Australopithecus afarensis of Aramis, ARA-VP-1-500, 3.8 Ma*

The ARA-VP 1/500 CT scan in exocranial basal view (Suwa *et al.* 2009) shows that the anterior end of the petrous pyramid was in contact with the SOS, the frontalization of the pyramids is a fact and therefore attests to an undeniable verticalization of the basi-sphenoid, while the semi-circular channels of simian type identify the fossil as an *Australopithecus* and not *Homo* (Figure 6.22).

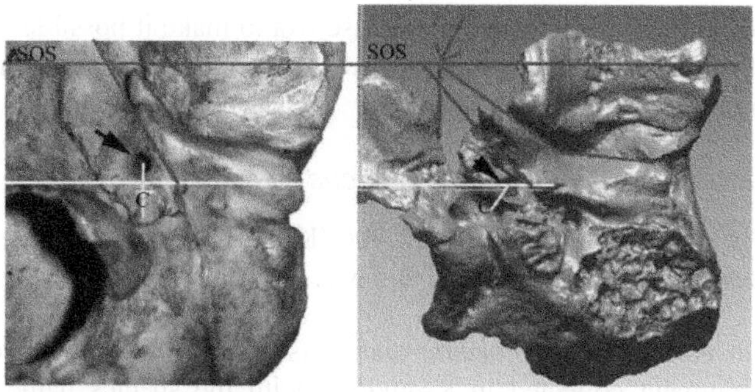

Figure 6.22. *Comparison of the orientations of the petrous pyramid between the gorilla and ARA-VP 1/500 aligned on the tangent to the basion. The carotid canal hole (black arrow) is on the tangent, as predicted by Dean and Wood's protocol. C: center of the jugular foramen. We notice the pivoting of the axis passing through the center of this foramen and that of the carotid foramen (yellow straight line), due to the advance of the clivus and the frontalization of the pyramid. The red arrow indicates the contact between the pyramid and the reconstructed SOS in relation to the surface of the TMJ (photo of the gorilla: A. Dambricourt Malassé, ARA VP-1/500 according to Suwa* et al. *(2009), different scales). For a color version of this figure, see www.iste.co. uk/dambricourt/embryogeny.zip*

6.3.1.9. Kenyanthropus platyops, *an australopithecine or human straightening?*

The skull, dated between 3.5 and 3.2 million years ago, was collected by Maeve Leakey in 1999, on the western shore of Lake Turkana (Kenya). The fossil is fragmented and deformed. Its species name *platyops* refers to the flat conformation of the middle face, which is considered as an argument to bring it closer to the genus *Homo*. Fred Spoor has reconstructed it and confirmed the first assessments:

> The cranium KNM-WT 40000 from Lomekwi has been assigned to a new genus hominin and species, *Kenyanthropus platyops*. Based on a substantially larger comparative sample than previously available, the results of statistical analyses show that KNM-WT 40000 is indeed significantly different and falls outside the known range of variation of species of *Australopithecus* and *Paranthropus*, contemporary *Australopithecus afarensis* in particular. These results support the attribution of KNM-WT 40000 to a separate species and the notion that hominin taxonomic diversity in Africa extends back, well into the Middle Pliocene. (Spoor *et al.*, *Hominin diversity in the Middle Pliocene of eastern Africa: The maxilla of KNM-WT 40000*, 2010)

The reconstitution of the endocranial base would make it possible to know more about its affinities with the genus *Homo*, as well as the semi-circular canals according to their state of preservation.

6.3.1.10. *Paranthropus or Australopithecus robustus?*

The skulls in question are KNM-ER 406 (Kenya), OH 5 (Tanzania), SK 48 and DNH 7 (Dremolen) in South Africa, 2–1.5 Ma.

The South African genus *Paranthropus* was created by Robert Broom in 1937 for a skull fragment collected in Kromdraii, with the species name *robustus*. In 1948, a robust mandible collected at Swartkrans was attributed to another *crassidens* species, to which a skull with the complete base, SK 48, was added.

This base is therefore the reference to define that of the genus Paranthropus *and to make a clear distinction with* Australopithecus robustus, *such as KNM-WT 17000.*

In exocranial view, the clivus of OH5 (Olduvai) shows a slope break between the basi-sphenoid and the basi-occipital, which is flat (Figure 6.23(A) and 6.23(B)). This conformation is reminiscent of the embryonic stage before the alignment of the basi-occipital with the basi-sphenoid. It can be seen in endocranial view, thanks to the disassembled skull with a very straight basi-sphenoid and a gently sloping basi-occipital (Figure 6.23(C) and 6.23(D)).

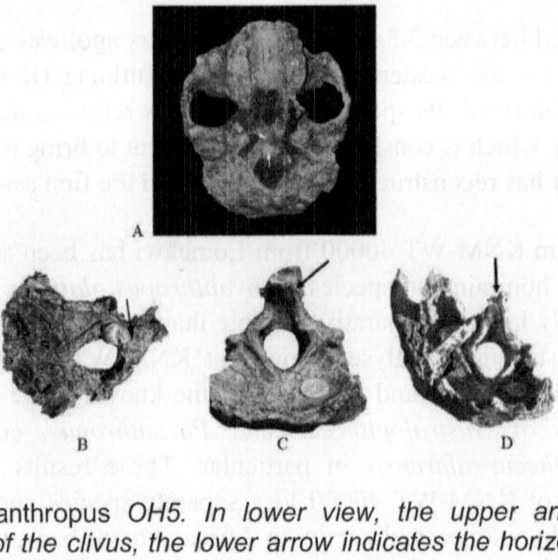

Figure 6.23. Paranthropus *OH5. In lower view, the upper arrow indicates the straightened part of the clivus, the lower arrow indicates the horizontal part. From B to D, the cast of the disassembled skull showing B: the exocranial face with the flat basi-occipital (black arrow) and above the very straightened part of the clivus, C: the endocranial view without the petrous pyramid and D: with the articulation of the pyramid and the basi-sphenoid, the arrows indicate the basi-sphenoid (photo: A. Dambricourt Malassé)*

This conformation is found on KNM-ER 406, with a very strong frontalization of the pyramids and incisor-canine alveoli (Figure 6.24).

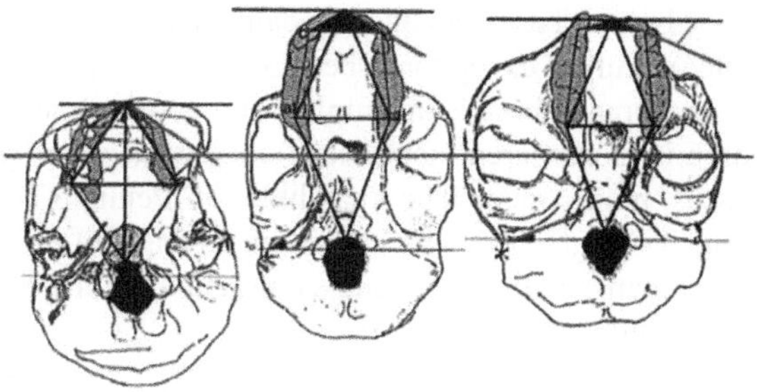

Figure 6.24. *Skulls drawn with a dioptograph, oriented in the Frankfurt plane, in lower view aligned with the post-orbital constriction that marks the anterior limit of the frontal lobes (transverse straight line red). From left to right* Homo erectus *(Sangiran 17 or* Pithecanthropus *VIII, Southeast Asia),* Australopithecus africanus *Sts5 and* Paranthropus boisei *KNM-ER 406 (Dambricourt Malassé 1999). For a color version of this figure, see www.iste.co.uk/dambricourt/embryogeny.zip*

The basi-sphenoid is very clearly verticalized and the frontalization of the pyramids is well correlated to its straightening and not to that of the basi-occipital (Figure 6.25). The craniofacial contraction angles of KNM-ER 406 are those of the juvenile Taung stage with a1 = 122° and a2 = 140°, the angle with the straightened segment and prosthion is 80°, which is the value of Taung's internal contraction with the prosthion.

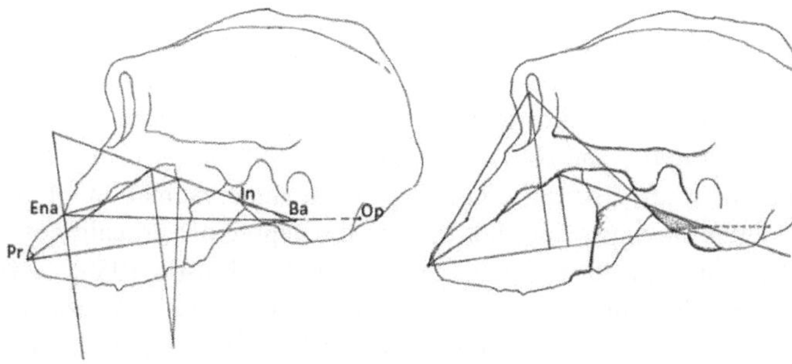

A B

Figure 6.25. *The angles of exocranial craniofacial contraction of KNM-ER 406, left from the basion (integrating the basi-occipital) and right with the straightened segment of the clivus. Ba: basion, Ena: anterior nasal spine, In: point of the maximum inflection of the exocranial base, Op: opisthion, P: prosthion (Dambricourt Malassé 1999)*

The single CT scan of the fossil (Figure 3 in Spoor *et al.* 2000) confirms this difference in the two-step straightening of the chordal chondrocranium. This morphology of embryonic origin is neither human nor australopithecine. The SK 48 skull is compacted, but its base remains typically that of the *Paranthropus*. The Dremolen (Eurydice) skull is exceptional because it is complete with its mandible and teeth, but the endocranial base is incomplete and poorly described. The shape of the mandibular arch is more flared than that of an *Australopithecus* and the basilar angular deviation is more open. The mandible of Peninj (Tanzania) shows a deviation of 80° and a straightened symphysis with an angle of 96°. These values are close to those of humans, with the strong contraction of the basi-sphenoid as a common embryonic context.

The three skulls ER 406 in Tanzania, OH 5 in Kenya and SK 48 in South Africa have this strong straightening of the basi-sphenoid in common, visible around 2 Ma or even less. Older skulls, such as Sterkfontein 5 and the "Black Skull" KNM-WT 17000, do not have such a pronounced straightening, nor this frontalization of the pyramids and incisor-canine alveoli. *Paranthropus* is therefore not an *Australopithecus robustus*. Its base confirms that the axial straightening of the *Hominidae* is the consequence of an amplification of the contraction of the fourth ventricle and has no relation to the much more limited telencephalization than that of a primitive species of *Homo*. The craniofacial contraction is different between the two genera: the incisor-canine occlusion is still modified, it is end-to-end with small crowns, while the molars and premolars are particularly developed. Cerebellization and new connections between the cerebellum and the brain have necessarily developed.

6.3.1.11. A skull of Homo habilis in South Africa?

The Sts 19 skull collected in Sterkfontein's Member 4 (2.6–2 Ma) has caused controversy. The cerebellar fossa shows strong affinities with *Homo sapiens* due to the straightening of the pyramids (Figure 6.26) and has therefore been attributed to the genus *Homo* (Kimble and Rak 1993). Ahern (1998) revised this diagnosis and concluded that it is an *Australopithecus* with an underestimation of the variety. However, the latter diagnosis does not include the degree of verticality of the clivus, nor that of the more upright pyramids (Broom and Robinson 1950; Clarke 1977). Since these are the first embryonic territories in which the basi-sphenoidal dynamics diverge very early between *Australopithecus*, *Paranthropus* and *Homo*, it is not possible to conclude that variety exists without having previously integrated the fact that the process of straightening and its amplitude are well-distanced embryonic regulations. This cannot be a variety because the changes concern embryonic organization. The study of the semi-circular canals made it possible to determine the taxon. If the canals turned out to be human, the skull is classified in the genus *Homo*.

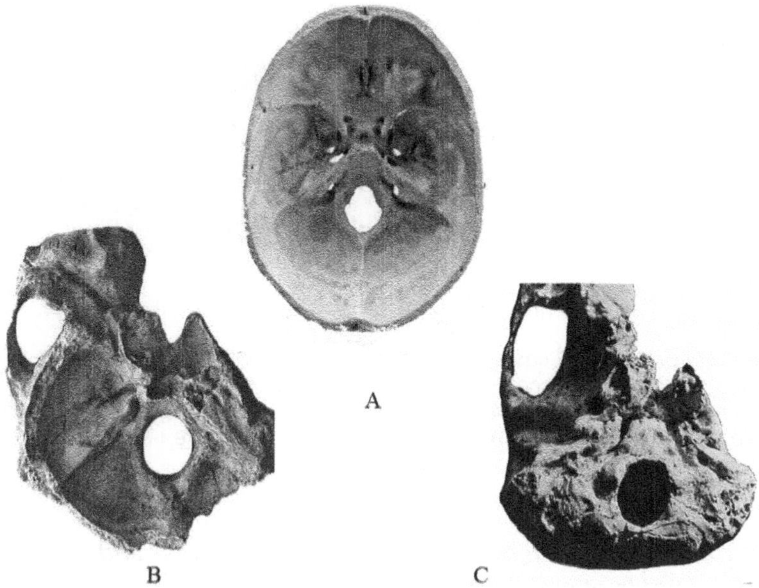

Figure 6.26. *Sts 19 skull compared with* Homo sapiens *(A). B: endocranial view and C: inferior view of the skull base (photo: A. Dambricourt Malassé, different scales)*

This debate is similar to that of another Sterkfontein fossil, the skull Stw 53 discovered in 1976 in Member 5, dated 1.8–1.5 Ma and attributed to the taxon *Homo habilis* (Curnoe and Tobias 2006) then *Homo gautengensis* (Curnoe 2010).

6.3.2. *Basi-cranial straightening, cerebellization and encephalization*

Dr. Charles Kimberlin Brain discovered a natural endocranial cast in 1966 in the Sterkfontein Cave, and Phillip Tobias commissioned Ralph Holloway to study it (Holloway 1972). This casting, SK 1585, is attributed to *Paranthropus robustus*. The Institute of Human Paleontology has a cast of good quality that can be compared to that of a gorilla. The difficulty for isolated endocranial casts is the definition of a reference plane. The horizontal plane in this case is given by the floor of the frontal lobes above the orbital cavities. The comparison clearly shows the winding of the occipital lobes downward, and that of the cerebellum and venous sinus forward, the whole following the direction of embryonic rotation of the sphenoid (Figure 6.27).

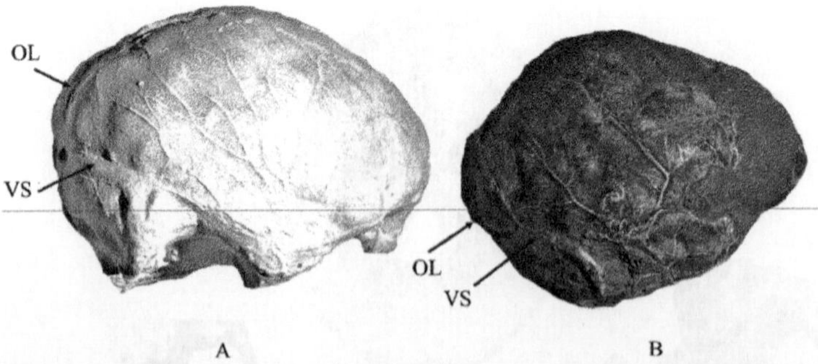

Figure 6.27. *Endocranial casts of a gorilla and Sterkfontein's* Paranthropus robustus *(SK 1585) at the same scale, showing the downward and forward curling of the occipital lobe (OL) and cerebellum with the venous sinus (VS) (photo: A. Dambricourt Malassé, same scale). For a color version of this figure, see www.iste.co.uk/dambricourt/embryogeny.zip*

6.3.3. Conclusion: Australopithecus, Paranthropus and prae-Homo, three thresholds of embryonic verticality

Paranthropus is no more than 2 million years old and it differs from *Australopithecus*, in that it presents a larger contraction of the fourth ventricle, not followed by the alignment of the basi-occipital, characteristic of the *Australopithecus*. Spoor also observed the conservation of the simian conformation of the semi-circular canals in *Paranthropus*, like *Australopithecus*. The post-cranial axial and appendicular skeleton are not known with certainty; recent collections have found an adaptation to arboreal living. The geological age of *Paranthropus* supports its filiation with *Australopithecus*. The neural kinetics of *Australopithecus* have thus evolved by showing embryos with a particularly vertical sphenoid body; however, the post-natal growth trajectories of the face have accentuated the simian component still visible on *Australopithecus*, namely, the face being carried toward the front of the cerebral skull. The adult orbits and the palate are no longer under the frontal lobes as in the infantile stage, but on the contrary are completely exocranial, while the development of the masticatory muscles results in a large fossa to house the temporal muscle between the zygomatic arch and the temples. It is further accentuated by the position of the cheekbones at the level of the center of the orbits, which is reminiscent of the *Lufgenpithecus* of Yunnan (Figure 6.4).

The face of the genus *Homo* does not follow this morphogenesis and does not stray far from the neurocranium. Even if prognathism is still apparent, underlined by the strong pneumatization of the frontal sinuses, a portion of the adult palate remains under the neurocranium (Figure 6.24). A new change of the regulation of the homeotic genes has thus altered the neural tube tensions expected from *Australopithecus* beyond the geometric limits of *Homo*, but without changes of the semi-circular canals; the result was *Paranthropus*.

The basic cranial conformation of *Paranthropus* shows how the embryogenesis of *Australopithecus* evolved. The basi-sphenoid contracted, but the basi-occipital was not straightened and the semi-circular canals remained simian. The gametes of *Australopithecus* continued to evolve, but the chondrocranial contraction differs from that of *Homo*. *Homo* embryogenesis developed from a population of great apes distinct from the ancestral strain *Australopithecus*.

The statistical study of mandibles and skull bases, both metric and anatomical, thus confirms that bipedalism was never the cause of the transformation of great apes into Hominids. Neither life choices nor subsistence behavior were the mechanisms of the straightening of the axial endoskeleton. The embryogenesis of the neural tube and the migrations of neuroblasts, the kinetics of the synchondroses of the chondrocranium and the regulation of the homeotic genes must be the first biochemical and physical processes to be considered. These kinetics and genetic regulations determined the future dynamic equilibrium between the face, the base, the spine and the appendicular skeleton, well before fetal growth. They were part of a complexification of the cerebrocerebellar neuron networks necessary for the psychomotor control of a central nervous system verticalized from the center of the base of the skull, and thus took longer to structure themselves. These morphogenetic processes which construct the embryo, the fetus, the newborn and then the child, who learns to walk, are to be placed in the continuity of gametogenesis, which is the cellular "space-time" of the mechanisms of evolution.

Its principles can be inferred from its effects: they are productive and self-regulated processes of complexification that are channeled to the nervous system, according to the cephalo-caudal gradient of the embryo and to regulations by the homeotic genes that manage the changes of the morphogenesis of the axial and appendicular skeleton.

Australopithecus is therefore not a genus of great ape, because it is no longer a great ape. Its level of organization is the evolution of this embryonic organization still living with the *paninea* and *ponginea*. It was a new "fundamental ontogenesis" (Dambricourt Malassé *et al.* 1999) with a diversification of species much more limited in time and space, since it ceased 3 million years after its emergence; *Paranthropus* is even more evasive.

The Linnaean notion of the family of *Hominidae* was created through a Fixist vision of the animal kingdom and has no ontogenetic reality, because there is not a single *Hominidae* embryogenesis. There are different embryogeneses, whose straightening of the dorsal cord reached verticality with *Australopithecus*, *Homo* and *Paranthropus*. The name *Hominidae* becomes a category constructed by the nomenclature to gather fossils whose axial endoskeleton is verticalized, but it is not representative of the mechanisms of evolution.

6.3.4. *Humanity's territory of otherness*

The consequences of a complexification of neural embryogenesis are therefore heterochronic, with a lengthening of the learning period and psychomotor development. The important psycho-affective relations between the infant and the mother are taken into consideration to a greater degree, as well as the relationships consciously reflected between the members of the same clan, particularly during critical situations. Indeed, a pregnant mother, or one raising her infant, could no longer flee by running on her four limbs, with her child clinging under her belly or on her back; she would carry it in her arms, but running in these conditions is almost impossible.

The periods of "mutagenesis" must have taken place as follows. Several clans were involved in this phase of the rapid complexification of gametogenesis; externally, the newborn "mutants" and "non-mutants" did not differ. The differences appeared progressively with growth. While some could still move quadrupedally on the ground, others remained bipedal and showed a delay in psychomotor development; these babies were more vulnerable and required more attention from their mothers. This did not prevent them from climbing trees, but they did so later, and their teeth were less aggressive as fangs disappeared.

Several scenarios are possible to explain the success of these critical phases for the mother-infant couple. The mothers of the same clan, sought after for a longer period of time by their very young children, were likely brought closer together, while reproduction was a rapid speciation barrier due to miscarriage, as the male and female gametes were only fertile in the same embryogenesis. Sentient behavior was therefore central to justifying the survival of the most vulnerable, and we agree with Albert Gaudry in his warning against the so-called explanation of natural selection as the cause of the success of any and all evolution.

The critical phase of hominization at the threshold of verticality depended especially on conscious reflexive and emotional behaviors, such as the ability to put oneself in the other person's place, to have empathy and to organize accordingly. This reflective ability of the central nervous system develops a sensory plasticity where the "sense of otherness" is structured. This is the reason why it is advisable to distinguish the hominization of the body and the humanization which engenders psychomotor and psycho-affective development at the time of the construction of the image of oneself and of the other, through various modes of communication, be they corporal or by vocalizations. This phase of humanization founds the meaning of the words "humanity" or "otherness". The subsistence and reproductive behaviors of mothers took place in densely wooded micro-habitats, but the biological territory of the hominization of the body was the placenta, generated by embryogenesis; this is the territory of the embryo. Meanwhile, the territory of humanization is a mother's struggle to recognize the infant as another self; otherness becomes the territory of humanity, in the neurosensory and, ultimately, most embodied sense (Dambricourt Malassé 2014). As for the emergence cradle of *Australopithecus*, the area stretching from Chad (Koro Toro, 3.3–3 Ma) to South Africa, on the one hand, and the conditions of both fossilization and exhumation of fossils in acidic (forest) environments, on the other hand, leave the question open on a continental scale.

7

Homo and Sapiens Embryogenesis

7.1. Fossil taxa of the genus *Homo*

Fossils of faces and skull bases attributed to extinct species of the genus *Homo* share anatomical, morphological and morphogenetic constants. There are enough mandible specimens to identify this morphogenesis and to place it in the evolution of embryonic straightening, no matter the volume of the cerebral and cerebellar hemispheres. The cranial bases are more common than those of the Australopithecines and they already attest – from an exocranial view – to a different spheno-occipital dynamic, with the alignment of the basi-occipital on the basi-sphenoid. The number of measured specimens is lower than the number of fossils observed since not all mandibles are measurable, sometimes the two mandibular bodies are brought closer together by the weight of the sediment, and thus the shape is significantly modified. Anatomy always complements biometry. Table 7.1 gives details of mandibles studied from Asia, North Africa and Europe, originally attributed to geographical varieties of *Homo erectus* and *Homo heidelbergensis* (Mauer).

In addition to these mandibles, there are Neanderthals from Europe and Asia, including young individuals (Table 7.2). Among the specimens from western Asia (Middle East), some were originally attributed to Neanderthals when they were *Homo sapiens* and vice versa.

Specimen	Country	l	L	D	α	β
Meganthropus 1952	Java	43	67	94	90	84
Pithecanthropus B	Java	31	60	74	76	110
Sinanthropus A2	China Beijing	30	59	–	–	–
Sinanthropus B1 phase 2	China Beijing	28	45	65	92	124
Sinanthropus G1	China Beijing	40	56	72	80	110
Sinanthropus H1	China Beijing	35	57	74	80	115
Sinanthropus K1	China Beijing	33	54	65	74	117
Atlanthropus I	Algeria	37	57	80	89	121
Atlanthropus II	Algeria	35	57	–	–	116
Atlanthropus III	Algeria	38	74	93	78	117
Rabat-Kebibat phase 3	Morocco	37	52	67	80	106
Mauer	Germany	32	68	80	72	100
Ehringsdorf phase 3	Germany	30	58	70	76	110
Montmaurin	France	29	64	75	72	110
Arago II	France	30	66	80	72	117
Arago XIII	France	33	75	91	75	117
Atapuerca	Spain	33	64	84	82	110
Banolas	Spain	31	70	85	75	100

Table 7.1. *Biometry of the mandible: extinct* Homo *taxa*

Since then, the taxa have become richer, with *Homo antecessor* (Atapuerca), *Homo georgicus* (Georgia) and *Homo denisovensis* (Siberia) in continental Eurasia, then *Homo florensiensis* (Java) and *Homo luzonensis* (Philippines) in insular Asia. In Africa, *Homo habilis*, *Homo ergaster* and *Homo rudolfensis* were supplemented by *Homo naledi* and *Homo gautengensis* in South Africa.

Specimen	Country	l	L	D	α	β
Krapina E	Croatia	34	70	85	75	105
Krapina G	Croatia	32	61	80	82	110
Krapina H	Croatia	41	–	–	–	107
Krapina J	Croatia	42	70	84	74	110
Vindija	Croatia	34	60	77	80	100
Ochoz	Slovakia	35	51	70	87	98
Monte Circeo III	Italy	35	71	(80)	–	95
La Chapelle-aux-Saints	France	35	70	(77)	–	105
Ferrassie 1	France	40	69	82	73	100
Hortus IV	France	36	57	80	85	100
Arcy-sur-Cure	France	40	67	80	73	100
La Quina H5	France	–	70	–	–	100
La Quina H9	France	38	58	70	74	94
Malarnaud	France	25	57	72	78	110
Petit Puy Medium	France	28	55	75	86	98
Regourdou	France	36	60	73	75	110
Spy 1	Belgium	37	74	86	71	103
La Naulette	Belgium	31	56	76	85	110
Amud	Israel	40	63	90	(91)	102
Pech de l'Azé phase 1	France	20	40	58	(93)	100
Roc de Marsal phase 1	France	21	42	60	(91)	105
Skhul 1 phase 1 Recons	Israel	20	45	66	(94)	90
Gibraltar phase 1	Gibraltar	22	46	62	85	109
Le Moustier phase 3	France	31	54	75	88	101
Qafzeh 9 Sapiens	Israel	39	69	(70)	–	82
Skhul IV Sapiens	Israel	43	64	75	72	97

Table 7.2. *Biometry of Neanderthal and proto-Cro-Magnoids (Sapiens) mandibles*

All of these mandibles have in common a large basilar angular deviation which is part of the variability of *Homo sapiens*, and an adult symphyseal angle always greater than 90° which is part of the variability of Australopithecines. The prognathism of the incisor and canine arches develops with growth, as shown by the five children; the symphysis tilts forward and downward, the angle opens. Variation in the size of the molars between taxa changes the length of the alveolar arch, its shape is therefore more or less parabolic. But what systematically distinguishes these mandibles from the current *Homo sapiens* is the absence of a chin, or *mentum osseum*, which is triangular in shape. The ossified symphyseal void, or *tuber symphyseos* – possibly sensitive to touch – forms a slight tuberosity in some Neanderthals, such as La Quina H9, but there are no lateral *fossae mentales* as in Sapiens. The oldest human mandible undoubtedly dates from 2.4 Ma. It was collected in Malawi (Uraha), south of Tanzania. A fragment of mandible with the symphysis, collected at Ledi-Geraru, Ethiopia and dated between 2.8 and 2.7 Ma, is attributed to *Homo*, although its symphysis is that of an *Australopithecus* with a very pronounced *alveolar planum* (Villmoare *et al.* 2015). The dental characteristics that bring it closer to *Homo* are related to diet, which is not a sufficient argument given a species of *Australopithecus* was able to adapt to various diets, depending on the ecosystem, and not reduce consumption to fruits, insects and nuts. In any case, it was not a post-natal diet that changed an *Australopithecus* face into a human one.

The lithic industry of Lomekwi (Kenya) dating from 3.3 Ma, and associated with faunal assemblages, argues in favor of meat consumption. The *Hominidae* behind these lithic industries was most probably *Kenyanthropus* (Harmand *et al.* 2015). Animal consumption has been evidenced by traces of cuts on two bones discovered in Dikika (Ethiopia) and dated at 3.4 Ma (McPherron *et al.* 2010). Therefore, the *Australopithecus* species would also have been able to adapt to a change in diet richer in animal protein and fat by consuming carrion and marrow.

The Ledi-Geraru mandible was reconstructed using symmetry, and so the basilar angular deviation could be measured. If the angle is close to 68° the fossil would be an *Australopithecus* species, if it is close to 78° it would instead be a *Homo* species. But in no case could a change in post-natal diet alter an *Australopithecus* face into a *Homo* face; either the embryogenesis of the neural tube generated tension that closed the IV ventricle and contracted the face in a similar way to *Australopithecus* forms, or the embryogenesis was that of *Homo*. *This fossil demonstrates the need to distinguish between the acquired structure and the embryonic structure, function and dietary adaptation.*

Finally, the occlusion of the oldest human arches can be seen in an adult *Homo georgicus* skull (1.8 Ma) and an adolescent *Homo ergaster* skull (1.6 Ma). In the latter, the lower canine is no longer in front of its upper counterpart but behind (psalidodontia), in the adult it is end-to-end (labidontia). In the Neanderthal, the occlusion of the temporary dentition is also like that of the adolescent *Homo ergaster*, and it becomes end-to-end with growth.

7.2. Fossil mandibles from *Homo sapiens*

To be sure that the oldest mandibles belong to *Homo sapiens* – i.e. the embryogenesis of present-day humanity, *Homo sapiens*, Linnaeus, 1758, or the Anatomically Modern Human (AMH) – it is preferable that the mandible is found in association with the neurocranium. This is the case with the numerous skeletons extracted from the caves of Skhul and Qafzeh in Israel, a few kilometers from the Mediterranean shore, formerly described as "proto-Cro-Magnoids". The adult neurocrania from Skhul are the oldest, dating back 115,000 years. They are complete and show a cerebellar fossa that is just as tilted down and forward as in *Homo sapiens.* The straightening of the basi-sphenoid and the basi-occipital is identical, with a sphenoidal angle of 115°. All the adult mandibles have a chin, their occlusion is still end-to-end and the dental arch is still prognathic, i.e. in front of the base of the symphysis, but the chin is present. This prognathism of the alveolar part is found in the maxilla, with the orbital cavities protruding out of the cranial fossa as a result of significant expansion (pneumatization) of the eyebrow arches. The skeletons from Qafzeh are 95,000 years old and share the same robust facial features, with prominent superciliary arches and an end-to-end occlusion. This forward development of the face shows the conservation of growth factors present in all extinct taxa, *Homo habilis*, *Homo ergaster*, *Homo rudolfensis*, *Homo rhodesiensis* in Africa, *Homo erectus* and *Homo florensiensis* in East Asia, *Homo georgicus*, *Homo antecessor*, *Homo heidelbergensis* and *Homo neanderthalensis* in West Asia and Europe.

Apart from those from Skhul and Qafzeh, the mandibles of *Homo sapiens* fossils come from Europe and date from the Upper Paleolithic (n = 37) and Neolithic (n = 57) as well as the Epipaleolithic in North Africa (Temara, Taforalt, Morocco, 17,000 years) (n = 21), giving a total of 115 specimens.

Specimen	l	L	β
Grimaldi	30	57	87
Combe Capelle	36	67	88
Cro-Magnon 1	37	80	79
Cro-Magnon 14	33	78	77
Solutré 1	23	51	82
Solutré 3	33	71	80
Solutré 5	30	55	80
Predmost 1	39	60	81
Predmost 2	27	61	87
Oberkassel 1	35	75	72
Oberkassel 2	30	65	84
Brno III	27	61	78
Kostenki XIV	30	63	87
Chancelade	37	70	68
Dolni Vestonice	26	60	71
Le Placard	24	64	69
Saint Germain	29	70	68
Saint Germain B1	24	60	87
Isturitz 1	36	69	76

Table 7.3. *Sample of mandibles from* Homo sapiens *of the Upper Paleolithic in Europe showing height of the symphysis (l), the length of the mandibular body (L) and the inclination of the symphysis (β)*

Systematic differences between the *Sapiens* mandible and other taxa are manifested at the symphysis. A *Sapiens* mandible always has a chin, protruding to a greater or lesser degree, the *tuber symphyseos* is surrounded by two *fossae mentales* and the symphysis angle, which is 90° at birth, closes instead of opening.

Can we establish a correlation between the presence of the chin and the embryonic degree of the basi-sphenoid rotation? Is the basi-cranio-facial embryogenesis of *Homo sapiens* the same as the oldest species that defines the *Homo* genus? Did the Neanderthal embryo develop according to the same neural dynamics as *Homo sapiens*? Was its basi-sphenoid as straightened as *Homo sapiens*? Was its midface as unprognathic at birth? Even more simply, what was the position of the cerebellum and the degree of straightening of the petrous pyramids of all of these African, European and Asian taxa? Is it consistent to give the name *Sapiens* to a 200,000-year-old skull fossil from Central India whose cerebellar fossa is as high as a 1.8-million-year-old African species? Is it consistent to give the name *Homo* to species that have nothing in common with *Homo sapiens* embryogenesis? Is this a remnant of fixist reasoning through nesting (Russian doll principle) and its evolutionary derivative, gradualism, which does not conceptualize emerging reorganizations?

Biological rigor leaves no doubt as to the criteria for attributing a skull to *Homo sapiens*. The rotation of the basi-sphenoid is the first embryonic movement that breaks the horizontal axis of the base. Its amplitude, followed by the rotation of the basi-occipital, determines the orientation of the brain stem and the spinal cord and changes the position of Meckel's cartilage with stresses in its ventral end. It is therefore biological nonsense to give the name *Sapiens* to skulls that have neither the embryonic rotation of the basi-sphenoid nor the straightening of the petrous pyramids of *Homo sapiens* Linnaeus, 1758, regardless of their geological age and cranial capacity.

7.3. The Neanderthal is not a *Sapiens*

The mandible of the Neanderthals is characteristic. The dental arch is displaced in front of the ascending branch which articulates with the cranial base, leaving a space between the last molar and the branch (retro-molar space), visible at all stages of dentition (Figure 7.1). Furthermore, the canine is on the side of the lateral incisor, so that the incisor–canine row forms a line rather than an arch. The imprints formed by the digastric muscles, inserted on the lower anterior edge of the mandible, are very pronounced; they crush the basilar margin so it becomes and widened. A young Neanderthal mandible cannot be confused with that of a *Sapiens*.

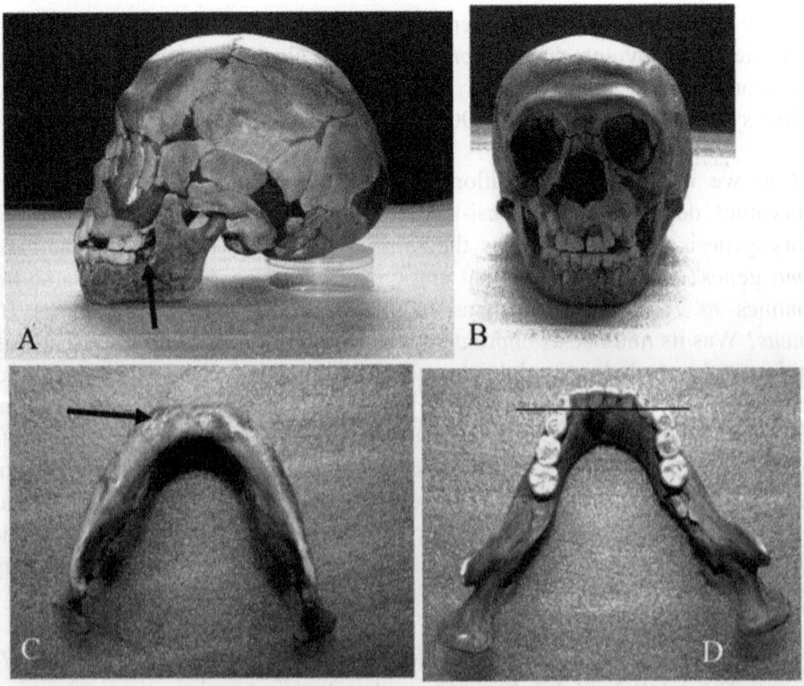

Figure 7.1. *The Neanderthal skull from Teshik Tash (child). A: with the retromolar space (arrow), B: the labidontic occlusion, C: the digastric imprints on the anterior pars basilaris (arrow) and D: the alignment of the canines with the incisors (photos: A. Dambricourt Malassé). For a color version of this figure, see www.iste.co.uk/dambricourt/embryogeny.zip*

7.4. The cerebellum position of extinct *Homo* species

The application of endocranial measurement protocols requires conditions that are rarely met. The bases must be complete, the fossils must have been scanned in their country of origin (France, in the case of this present research program) and be accessible to researchers. Exocranial protocols on casts have confirmed that the mandibles of all these extinct species belong to the same morphogenetic unit, regardless of the volume of the cerebral hemispheres and cerebellum, and independently of the geographical diversity that distinguish them by the shape of the molars or semi-circular canals, for example. All are part of cerebellar fossa rotation that has never exceeded the amplitude of the oldest known skulls and their contemporaries collected in Western Asia (Georgia) and Africa (Figure 7.2).

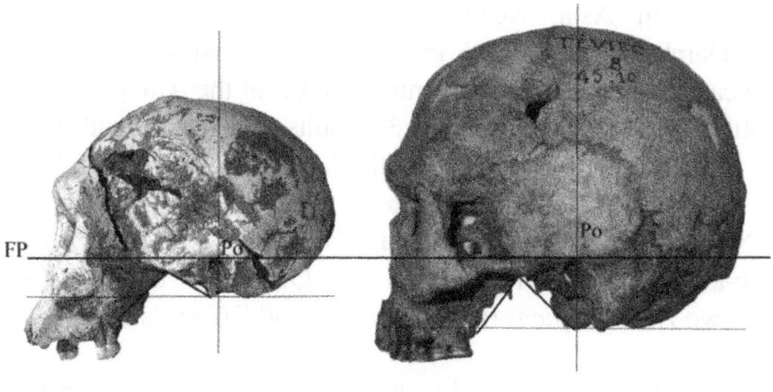

Figure 7.2. *Skulls, A: Homo georgicus (1.8 Ma) and B: Homo sapiens (Teviec, 7,400 years BP) in the Frankfurt plane and centered on the porion comparing the positions of the cerebellar fossa and basi-cranial contraction (triangle) (photos: A. Dambricourt Malassé). For a color version of this figure, see www.iste.co.uk/dambricourt/embryogeny.zip*

The comparison of skulls oriented in the Frankfurt plane and centered on the porion allows an assessment to be made with the African *Homo ergaster* KNM-ER 3733 (1.9 Ma), *Homo habilis* KNM-ER 1813 (1.9 Ma), *Homo rudolfensis* KNM-ER 1470 (1.88 Ma), KNM-WT 15000 (adolescent, 1.8 Ma), then later skulls in southern Africa such as *Homo rhodesiensis* (400,000–300,000 years) and *Homo naledi* (335,000–236,000 years), and in northern Africa with *Homo sp.* (Jebel Irhoud 300,000 years), in Europe with *Homo antecessor* from Spain (800,000 years old), *Homo heidelbergensis* and all the Neanderthals, in Asia with *Homo georgicus* (Georgia, 1.8 Ma), *Homo erectus* from Java, China and the only fossil from the Indian subcontinent found in central India at Hathnora which is 200,000 years old.

The higher position of the cerebellar fossa corresponds to a much less pronounced inclination of the pyramids than in *Homo sapiens.* These differences were noted as early as Franz Weidenreich's first exhaustive comparisons with Sinanthropus (*Homo erectus* from Beijing) (Figure 2.2). The inclination of the petrous pyramids was not measured until 1970 for the study of the Neanderthal skull from Amud (Israel) by Suzuki (1970). This study showed angles in *Homo erectus* and *Homo neanderthalensis* that were always greater than 90°, whereas the value is less than this angle in *Homo sapiens* with tall, slender pyramids. The rotation of the basi-sphenoid which drives the pyramids is therefore less pronounced in extinct *Homo* taxa.

Angular measurement protocols (sphenoidal rotation, sphenoidal angle and pyramid straightening) were applied to CT scans of current and fossil *Homo sapiens,*

Homo erectus from Asia, Neanderthals from Europe and chimpanzees (*Pan troglodytes)* (Dambricourt Malassé *et al.* 2022). Measurements on current *Homo sapiens* were carried out with Dr. Fabienne Lallouet at the American Hospital in Paris, on a population of 30 adolescents and 30 adults, 50% male and 50% female. The chimpanzee collection included 41 skulls including 32 young individuals from 3 months to 9 years and 9 adults. The fossil skulls consisted of three *Homo erectus* from Java (Ngandong 7, Ngandong 12 and Ngawi 1), four *Homo neanderthalensis* (La Chapelle-aux-Saints, La Ferrassie 1, La Quina H5 and Le Pech de l'Azé), completed by two *Homo sapiens* (Cro-Magnon 1 and l'Abri Pataud).

The sphenoidal rotation angles of *Homo erectus* and Neanderthals were more open and corresponded to pyramid inclinations that were less straight than in Sapiens. Some of the skulls with an axial endoskeleton that is less vertical than Sapiens, could be distinguished by a dynamic in extension. These were European lineages with the Neanderthals, Asian lineages with *Homo erectus* from Java (Sangiran 17) and China (*Homo erectus pekinensis*), and North African lineages with Jebel Irhoud dating back 300,000 years. These bases have a less pronounced pyramidal inclination than older taxa. These differences with *Homo sapiens* are found in exocranial view with the higher position of the cerebellar fossa relative to the Frankfurt plane, the mastoid process always higher than in *Homo sapiens*. Figure 7.3 shows the example of extension from *Homo georgicus* (1.8 Ma) in Asia with Sangiran 17 (uncertain age, middle Pleistocene), in North Africa with *Homo sp.* (Jebel Irhoud, Morocco, 300 ka) and in Europe with *Homo neanderthalensis* (La Chapelle-aux-saints, 60 ka).

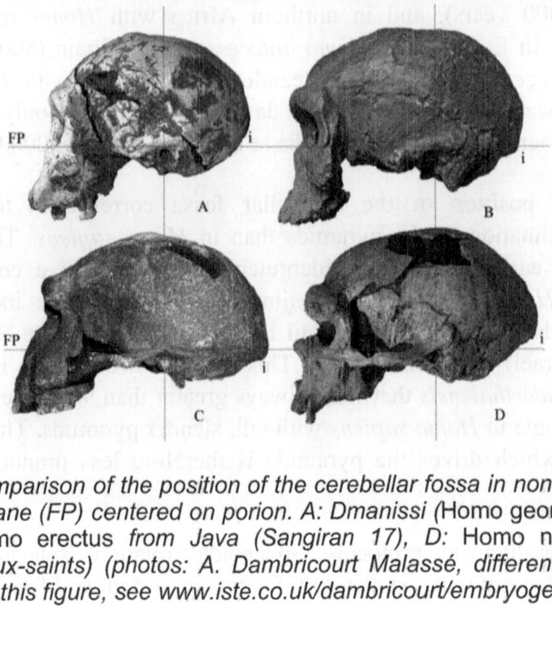

Figure 7.3. *Comparison of the position of the cerebellar fossa in non-sapiens taxa in the Frankfurt plane (FP) centered on porion. A: Dmanissi (*Homo georgicus*), B: Jebel Irhoud, C:* Homo erectus *from Java (Sangiran 17), D:* Homo neanderthalensis *(La Chapelle-aux-saints) (photos: A. Dambricourt Malassé, different scales). For a color version of this figure, see www.iste.co.uk/dambricourt/embryogeny.zip*

When the skulls are centered on the porion, close to the axis of rotation, an elongation thus appears forward of the porion from three continental areas (Northwestern Africa, Western Europe, Southeastern Asia) while the cranio-spinal field does not lower, as evidenced by the high position of the mastoid process.

The Neanderthal's stages of development, in profile view, show the accentuation of the extension process with less straightened pyramids from the earliest stages, such as the Pech de l'Azé (2–3 years old) (Figures 7.4 and 7.5).

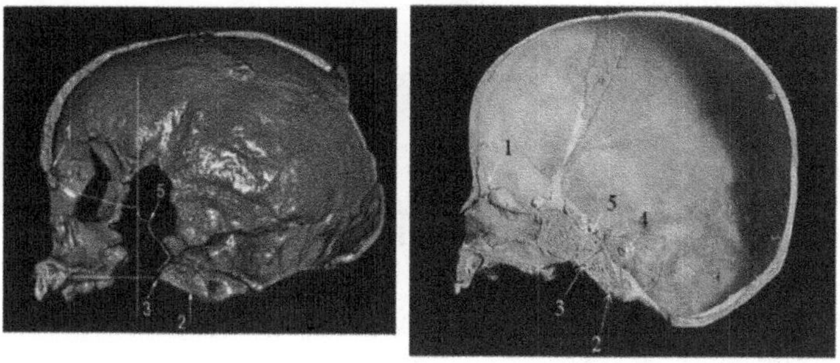

A B

Figure 7.4. *CT scan of the Pech de l'Azé Neanderthal child (A), in right lateral endocranial view. 2: basion, 3: location of the spheno-occipital synchondrosis, 4: internal auditory meatus on the petrous pyramid, which shows the slightest straightening in comparison with Sapiens at the same dental stage (B) (photos: A. Dambricourt Malassé). For a color version of this figure, see www.iste.co.uk/dambricourt/embryogeny.zip*

The occipital squamous remained in an elevated position, the cerebellum developed backward and not forward and downward.

The skeletal morphogenesis of Neanderthals is the best documented of the extinct species with some neonatal stages and very young children. The best preserved are Mezmaiskaya (northern Georgia, aged 1–2 weeks), Le Moustier II (southern France, aged 9 months) and two skulls from Dederiyeh (northern Syria, aged 19 and 24 months). The conservation of Mezmaiskaya (Ponce de Leon *et al.* 2008) and Le Moustier (Gunz *et al.* 2011) would allow comparison of the straightening of the clivus and pyramids with *Homo sapiens*, which is unique in the Neanderthal collections. The reconstruction of the Le Moustier has confirmed our conclusions that have been disseminated since 1987, that the craniofacial ontogeny of the Neanderthal differed from that of Sapiens as early as the intrauterine period, with prognathism already in place. The position of the body of the sphenoid, the clivus and the pyramids was said to be "subtle". If it had been "modern", the authors would have announced it, because this would have been the proof that Neanderthals belonged to the Sapiens species. This is a dogma that has prevailed in France for the last 40 years, but is falling into disuse when confronted with the principles of evolution: (1) embryogenesis is chronologically first in phylogenesis and (2) a species is formed after the subspecies stage and not the opposite (process of speciation). *The already prognathic midface of the Neanderthal newborn cannot be associated with such pronounced basi-sphenoid embryonic tensions as in Sapiens.* The vertical orientation of the petrous pyramids is predictable, it was less upright and this could be verified by the angle between the temporal and cerebellar faces of the petrous pyramid.

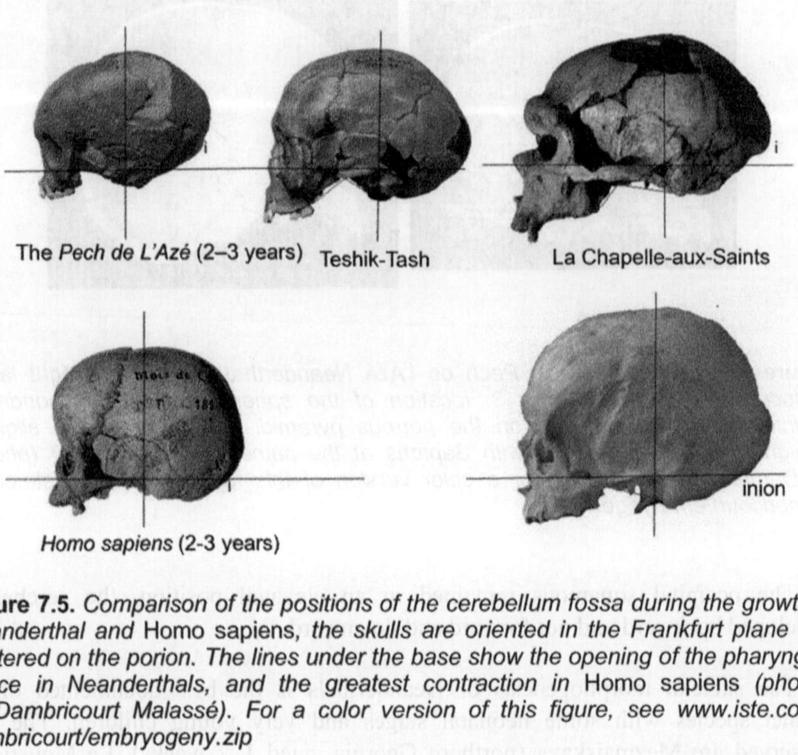

Figure 7.5. *Comparison of the positions of the cerebellum fossa during the growth of Neanderthal and* Homo sapiens, *the skulls are oriented in the Frankfurt plane and centered on the porion. The lines under the base show the opening of the pharyngeal space in Neanderthals, and the greatest contraction in* Homo sapiens *(photos: A. Dambricourt Malassé). For a color version of this figure, see www.iste.co.uk/dambricourt/embryogeny.zip*

The study of the cranio-palatal balance of La Chapelle-aux-Saints compared to Cro-Magnon reveals a balance of constraints between the base, the face and the occlusion, which is unrelated to *Homo sapiens* (Dambricourt and Deshayes 1995). The sphenoidal angle is 134°, very open, and the cerebellar fossa is in an elevated position. If we look at the nasal opening from the front, we see the axis vertebra in the background, whereas in *Homo sapiens* it is the basi-occipital. The Neanderthal does not have the cranio-palatal balance of *Homo sapiens*, its occluso-postural balance was quite different (Figure 7.6).

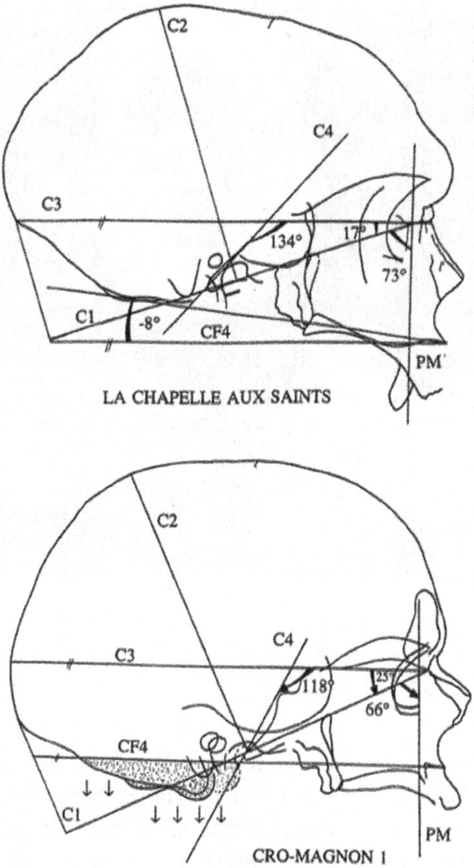

Figure 7.6. *Comparison of the cranio-palatal tracings of La Chapelle-aux-saints and Cro-Magnon (Dambricourt and Deshayes 1995). For a color version of this figure, see www.iste.co.uk/dambricourt/embryogeny.zip*

The extension of the face results in the swelling of the midface (oncognathism) which corresponds to significant pneumatization of the maxillary sinus (Heim 1989). The pneumatization is found as far as the supraorbital torus with the continuous passage between the two eyebrow arches. The alveolar-dental drawer is thus moved forward and a space is formed between the last tooth of the mandible and the rising branch. The extension is clearly seen in the medial sagittal section, with the pterygoid apophyses forming a bridge extending between the sphenoid and the palate (Figure 7.7).

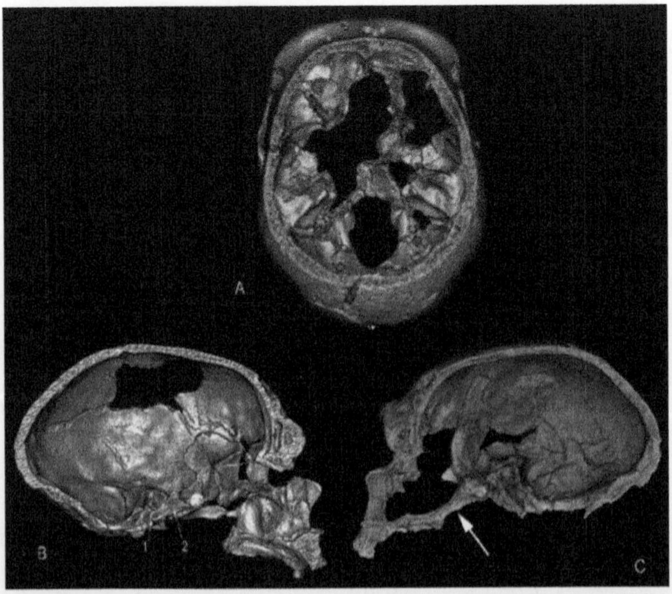

Figure 7.7. *CT scan of La Chapelle-aux-Saints (collection of the* Musée de l'Homme*). A: upper view, B: left lateral view, C: right lateral view, the arrow indicates the pterygoid process (photos: A. Dambricourt Malassé)*

The postcranial and appendicular axial skeleton of the Neanderthal is very similar to that of *Homo sapiens* and differs from it only in the conformation of the semi-circular canals, since the posterior is lower than in older taxa (Spoor *et al.* 2003). The shape of the bones differs – clavicle, scapula, pelvis, vertebrae, barrel-shaped rib cage, foot and hand bones – while the prefrontal cortex is less developed. Embryogenesis and regulation of the homeotic genes were quite distinct from *Homo sapiens*, the closure of the IV ventricle did not follow the same rhythm, the formation of the circular canals followed other apoptosis processes. Moreover, the conformation of their canals is not specific to them, as it is found in China with a 100,000-year-old skull found in Xujiayao, which is part of another lineage, as Wu Xijie of the Institute of Vertebrate Paleoanthropology and Paleontology (IVPP) in Beijing concludes (Wu *et al.* 2014).

Skulls that do not have the lowering of the cerebellar fossa of the anatomically modern man were simply not *Homo sapiens*, their occluso-postural balance was different. We have always concluded that the spinal bearing of Neanderthals differed from Sapiens due to a different supra-chordal neural embryogenesis with divergent dura-merian tensions (base extension). Reconstructions of vertebral posture, such as that of the Neanderthal skeleton of Kebara (Israel), confirm this, with less pronounced curvatures (Been *et al.* 2017), which restores some validation with respect to Marcellin Boule, even if he did exaggerate. *Everything fits together to demonstrate that the axial straightening and craniofacial contraction of the Neanderthal never exceeded the threshold of the first species which defines* Homo *and that Sapiens is the complexification and reorganization of this* Homo *embryogenesis. Homo sapiens* Linnaeus, 1758 is indeed the last threshold of embryonic straightening of the dorsal cord.

We have deduced that there must be cognitive specificities even in the development of language functions, the vocal apparatus being different, as well as meningeal vascularization (Saban 1995), in order to restore the specificities of Neanderthals (Dambricourt Malassé and Deshayes 1995) and we continue in this direction (Dambricourt and Joffroy in progress).

7.5. What is the relationship between the Neanderthals and the Sapiens?

7.5.1. *Skhul in Israel (Near East): the oldest Sapiens burial but with a young Neanderthal too*

The Skhul Cave is the oldest burial place known to date (115,000 years) with the skeletons of seven adults and three children attributed to Anatomically Modern Humans, or *Homo sapiens sensu stricto.* An incomplete skull of a child was extracted from a burial with a face and neurocranium that are "modern". On the other hand, the mandible attributed to him, with its complete set of deciduous teeth, is typically Neanderthal and does not fit with the oversized neurocranium (Figure 7.8).

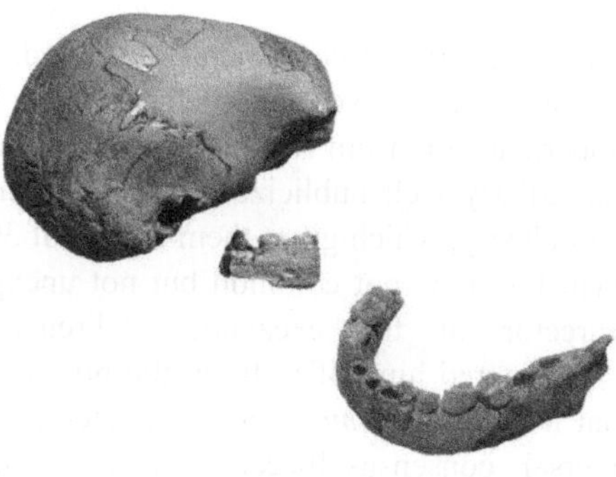

Figure 7.8. *The infantile Sapiens neurocranium, Skhul 1 and the mandible attributed to it but which is too small to match. The details show that the symphysis and the incisor–canine arch are Neanderthal features (photo: A. Dambricourt Malassé)*

After verification, it appeared that the mandible had been collected in a breccia among the excavated material (McCown and Keith 1939) and was therefore not actually associated with the child's skeleton. This cave is supposed to have never yielded any Neanderthal remains, but this mandible at least attests to the probable burial of a very young child. The consensus is that the presence of modern man in the Middle East predates the presence of Neanderthals from Eurasia. If this mandible was contemporary of burials, this would mean that Neanderthals occupied the Levant for at least 115,000 years; that would change the explanation for the presence of Neanderthal genes in the *Homo sapiens* genome, through gene flows with its ancestors such as Mugaret-el-Zuttiyeh (*Homo heidelbergensis*).

Neanderthals and *Homo sapiens* cohabited in the Levant, and interbreeding probably took place. The question of hybridization is a complex subject that does not address embryogenesis and, consequently, does not measure the difficulty of admitting the viability of embryos. The regulation of the homeotic genes was totally different, and the question arises as to whether male and female gametes from the two lineages could reasonably be expected to produce an embryo. Hybridization was possible between the ancestors of *Homo sapiens* and Neanderthals from Asia who would have crossed into the Levant because the embryogeneses were very close, which could justify the presence of some Neanderthal genes in *Homo sapiens* in Eurasia and in Africa. These genes did not concern the embryonic regulation and could be transmitted in the new Sapiens embryogenesis. But the viable hybridization between a derived *Homo* embryogenesis such as Neanderthal engaged in embryonic extensional tensions, and a reorganized embryogenesis with neural tube tensions more pronounced than that of the *Homo* genus, seems unrealistic to us.

7.5.2. *Jebel Irhoud in Morocco (North Africa) is not a* Homo sapiens

The fossils found at Jebel Irhoud correspond to 13 individuals and include a juvenile (phase 2) mandible. Their taxonomic determination has always generated debate between a Neanderthal form from North Africa and an archaic *Homo sapiens*. These fossils were particularly well publicized in 2017 when these remains were dated back to the lithic industry, which gives them an age of 300,000 years. Finding 300,000-year-old human fossils is not common but not unexpected either, so why the furore? The director of the excavations, French paleoanthropologist Jean-Jacques Hublin, considered himself to be in the presence of the oldest *Homo sapiens*, specifying that it was not *Homo sapiens* Linnaeus, 1758. This opinion did not meet with universal consensus because it was subjective. Intrauterine ontogenetic chronology was not taken into account as it had no phylogenetic basis. No anatomical feature matched that of *Sapiens*: mandibles without a chin, inclination of the petrous pyramids, position of the cerebellar fossa and imprints of the cerebral hemispheres.

The notion of archaic *Homo sapiens* was constructed within the framework of the gradualist evolutionary paradigm and without any embryonic consideration. An archaic form of *Homo sapiens* was defined in relation to current-day anatomy by following the chronology of development since embryogenesis. The qualifier "archaic" refers to the conservation of ancestral characters independent of the new embryogenesis. Middle Eastern proto-Cro-Magnoids according to the old expression, such as Qafzeh and Skhul, are "archaic" *Homo sapiens* because they still developed frontal sinuses and an end-to-end occlusion, whereas they were effectively Sapiens from the first movement of the neural tube which straightened the basi-sphenoid and the basi-occipital, tilting the otic capsules forward and downward.

The classification of Jebel Irhoud does not integrate the skull base and consequently embryogenesis is not conceived as the key to reading the first of the diagnoses.

The juvenile mandible from Jebel Irhoud also has no chin, and the *tuber symphyseos* is no better cleared than in the Neanderthal mandible of La Quina H9 (Figure 7.9).

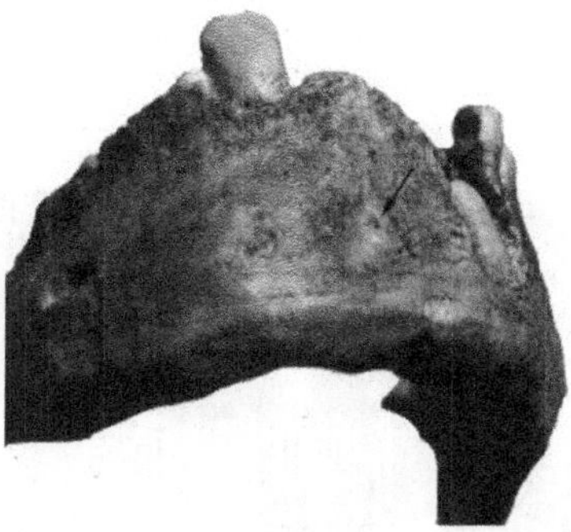

Figure 7.9. *Juvenile mandible from Jebel Irhoud (cast) with the tuber symphyseos (arrow) and absence of a chin (photo: A. Dambricourt Malassé)*

Similarly, the adult skull from Jebel Irhoud had not surpassed the straightening of the oldest *Homo*. The comparison with Qafzeh 6 and Cro-Magnon 1 shows mastoid apophyses in the upper position, a slope of the petrous pyramids that is much less straightened, and an extension of the occipital lobes toward the back that is more similar to a growth dynamic seen in Neanderthals and exactly the opposite of that expected for *Homo sapiens* (Figures 7.10 and 7.11).

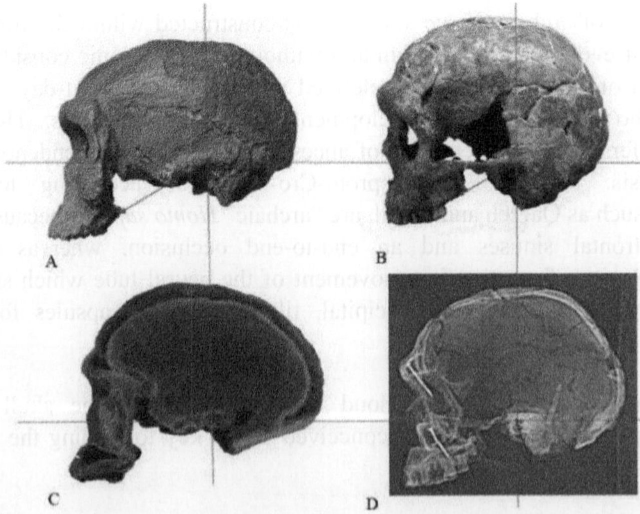

Figure 7.10. *Comparison of the cerebellar fossa position and basi-cranio-facial contraction between Jebel Irhoud (Morocco) and Qafzeh 6 (Israel). A and C: cast and CT scan of the cast of Jebel Irhoud, B and D: original skull of Qafzeh 6 and CT scan carried out by E. de Kerviler and J.-P. Martin in 1999 at the Saint-Louis Hospital in Paris (photos: A. Dambricourt Malassé). For a color version of this figure, see www.iste.co.uk/dambricourt/embryogeny.zip*

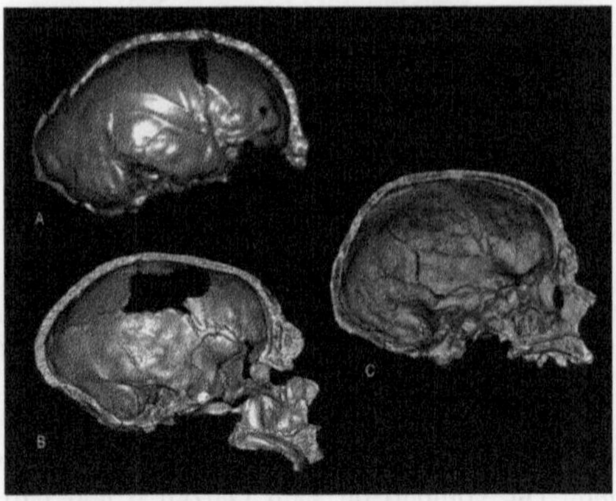

Figure 7.11. *Comparison of the left endocranium by CT scan, A: Jebel Irhoud (cast, CT scan from American Hospital of Paris), B: La Chapelle-aux-Saints and C: Cro-Magnon 1 (originals, CT scans from the* Muséum national d'Histoire naturelle. *The CT scans clearly show the less upright slope of the pyramid in Jebel Irhoud and the formation of an occipital bun (photos: A. Dambricourt Malassé). For a color version of this figure, see www.iste.co.uk/dambricourt/embryogeny.zip*

The embryonic constraints acting in sphenoidal synchondrosis ceased earlier than in Sapiens. And without maintaining these constraints, the brain could only grow in the zones of least resistance, in height and behind. The occipital lobe thus formed the famous occipital bun (Figure 7.11). *Consequently, Jebel Irhoud cannot be identified as a* Homo sapiens, *even archaic, nor was he a Neanderthal according to the shape of the mandibular arch and brain endocast.* As for seeing it as an ancestral lineage of *Homo sapiens*, it seems unlikely for two major reasons:

1) *the dynamics of the synchondroses at the base of the skull were already drifting in the direction of extension, unlike Broken Hill (*Homo rhodensiensis, *Rhodesia), which is almost contemporary; and*

2) *the first molar of the maxilla does not have the cusp pattern common to* Homo sapiens *and* Homo neanderthalensis.

This pattern is characterized by the pronounced development of the posterior and lingual cusp (hypoconus), which gives a diamond shape (rhomboid) to the occlusal surface. The pattern in Jebel Irhoud has a square occlusal surface because this cusp is poorly developed. It is also the pattern of an older Moroccan maxilla collected in Rabat (370,000 years old) and that of the oldest known species in Western Asia with *Homo georgicus* (1.8 Ma), and in Africa with *Homo habilis* (OH 24, KNM-ER 3733, KNM-ER-15000 and KNM-ER 62000). This morphology dates back to at least 1.8 Ma. On the other hand, the rhomboid shape is not known to have existed before 1 million years ago, with the oldest specimen collected in South Africa at Cornelia-Uitzoek. The rhomboid pattern is found in North Africa with specimens dating back at least 700,000 years found in Algeria (Tighennif, ex-Atlanthropus), in South Africa with *Homo naledi* dated at 250,000 years old and in South China dating back more than 200,000 years (Tongzi) (Dambricourt Malassé *et al.* 2018). Finally, this pattern is visible on the oldest skull of *Homo sapiens* Herto, dating back 160,000–154,000 years, collected in Ethiopia. The populations of Jebel Irhoud thus lack the morphogenetic characteristics expected from the neural embryogenesis of Sapiens, on the one hand, and from the ancestral lineage common to Sapiens and Neanderthals, on the other hand.

In the 21st century, the phylogenetic study of the head still ignores the skull base or chondrocranium. For example, the adult craniofacial morphotype of the common ancestor of *Homo sapiens* and *Homo neanderthalensis* has no base (Mounier and Mirazon Lahr 2019); most of the neurocranium that accounts for axial straightening and retrognathism is not considered. The ancestral morphotype of the cranial base is the one that defines the genus *Homo*. Jebel Irhoud is already drifting toward the extension, which suggests endemic effects and specialization due to a dead-end configuration of the occupied territory.

The research dedicated to *Homo sapiens* emergence is limited to the adult brain and face, without any doubt that gradualism will never be the right key to reading this, since this emergence concerns embryogenesis, its complexification and the cranio-sacral tensions whose modalities follow the threshold effects and are therefore saltationist. Discontinuity and threshold effects will remain intellectual obstacles for a long time to come because the absence of gradualism over short time scales is like the loss of a guiding thread. But if the problem of Sapiens emergence is placed in very long geological time scales, we can again find continuity, and it becomes that of a process of increasing complexity during gametogenesis, and the threshold effects are better assimilated by conceptual thought.

7.5.3. *Homo floresiensis, the Asian Hobbit, parent of Homo habilis*

Human fossils from the island of Flores, located in the Southeast Asian archipelago, were discovered in the Liang Bua Cave in 2003. The nickname "Hobbit" was given to them because of their height of 1 m, with a cranial volume of 140 ml. Currently dated at 80,000 years old, these remains have divided the anthropologist community because the first dating gave an age of 12,000 years, while Sapiens characteristics such as the chin are absent. However, the doctrine holds that *Homo sapiens*, in the strictest and most rigorous sense of embryonic biology, is the only *Hominidae* since the extinction of the Neanderthals around 28,000 years ago. The base of the endocranium, and therefore the degree of straightening of the nervous system, has not been studied but the high position of the cerebellum fossa and the receding mandibular symphysis demonstrated, without doubt, that these skeletons were not *Homo sapiens*. The consensus on being a close relative to *Homo erectus* was achieved not because of the anatomy but because of the new dating of 80,000 years, consistent with the absence of *Homo sapiens* in East Asia at that time. The comparisons caused surprise by demonstrating that this taxon was derived from a very ancient lineage of *Homo habilis*, which would have reached Asia long before *Homo erectus* and is relevant to the Masol discovery in India (see section 7.6).

7.5.4. *Conclusion: the embryonic threshold between Homo and Sapiens*

We can no longer define the *Homo* ancestral morphogenesis by ignoring the skull base which connects the brain stem to the spinal cord. It is not by chance that in 40 years of research it is generally the practitioners confronted with this reality of our anatomy verticalized from occlusion to coccyx, who have welcomed the discovery, preferring to concern themselves with predictions, instead of ignoring it on the grounds that it is in conformity with the orthogenesis of Teilhard de Chardin and not with the neo-Darwinian dogma.

Working to model the effects of a future craniofacial contraction threshold for public health purposes is a wise move. What does it matter if the neo-Darwinian dogma does not conceive the reproducibility of the processes; *Homo sapiens* is not located anywhere in the curve of increasing complexity-consciousness, which would claim that the process will not happen again, has no measure of the consequences, because verticality has reached its limit and the physiology as well as the psychomotor development would be modified with effects that do not correspond to the mental images we make of ourselves. These struggles against the orthogenesis of the hominization process do a disservice to our humanity (Lecointre 2012), overcoming them manifests the awareness of a collective responsibility, a threshold of maturity in the process of humanization since the Greek philosophers of Antiquity, in respect of the scientific approach and the awareness of its limits. Mathematical theory will never be evolution and researchers are not its "great architect", with the risk of opening the doors to minsinformed transhumanism. Physico-mathematicians are trying to approach the laws between entropy and negentropy and as René Thom so rightly put it, "predicting is not explaining" (Thom 2009).

In conclusion, there is no basi-cranio-spinal morphogenesis common to the extinct species of *Homo* and Sapiens, the embryonic angular discontinuity is radical because it is biological and physical. The taxon *Homo* is a fundamental ontogenesis probably as old as that of *Australopithecus*, with a great diversity of species or subspecies in Europe, Asia and Africa, with no change in the axial plane of organization in which the degree of basi-sphenoidal rotation occurs. Neural tube vesicles that are not directly in contact with the chord, such as the future cerebellum and cerebral hemispheres, have continued to evolve according to different encephalizations, while the dynamics of sphenoidal synchondroses have varied following extensional tendencies. These lineages have become extinct at different rates, while others were maintained, some of them reaching an optimal threshold of complexity in their gametogenesis. This threshold has clearly activated a genetic memory regulating increasing complexity, generating a reorganization and complexification of cranio-spinal neurogenesis and morphogenesis, thus a new fundamental ontogenesis called Sapiens has emerged. The focus of emergence remains hypothetical.

With the current state of knowledge, still dependent on the conditions of fossilization and exhumation of fossils, East Africa is favored as the cradle of emergence because of the Herto skull. The reality is more complex, in the sense of a probable "fertile crescent" of genetic mixing between East Africa and South Asia. Indeed, the oldest *Homo sapiens* skull from the Indian subcontinent, Orsang (nearly 50,000 years old, Chamyal *et al.* 2011), retains a very ancient characteristic of Asian *Homo erectus*, which is at least 1.8 million years old, and still visible on an encephalized *Homo erectus* in central India: the Narmada skull. This skull is not *Homo sapiens*, as Marie-Antoinette de Lumley has demonstrated, being well aware of the winding of the skull base and its consequences on the position of the zygomatic arch (de Lumley and Sonakia 1985). This very old characteristic is a small relief behind the ears called the *torus angularis*. It has never been observed in Africa, with the exception of Broken Hill, the only known and very late case of a visible contribution of Asian genes (Dambricourt Malassé *et al.* 2015). The genes of old Asian lineages thus circulated between southern Asia and eastern Africa and were found in ancestral lineages of Sapiens, which is why it is equally likely that the Arabian Plate was an emergence hotbed when the climate was wetter and the land was rich in plants and game.

7.6. The collapse of a paradigm: Man was present in Asia before the end of the Tertiary Era

The oldest cut marks of butchery activities (Figure 7.12) associated with the remnants of an operating chain were discovered in 2009 in the Himalayan foothills of northwest India, at Masol in Punjab (Dambricourt Malassé 2016, 2021a; Dambricourt Malassé *et al.* 2016a, 2016b). The sediments are silts deposited during floods of Himalayan rivers that carried away many herbivorous species and deposited quartzite pebbles from Himalayan terraces. The age of these markings, thought to be more than 2.6 Ma (Chapon Sao *et al.* 2016) and now known to be at least 2.8 Ma (Chapon Sao *et al.* 2018; Dambricourt Malassé 2021a), was a huge surprise because the paradigm did not envisage the first human settlements in Asia being before 2 Ma. The site, spread over 50 hectares, was an ancient tectonic dome that had been eroded by monsoon rains. It is exceptional because it is the oldest site which, even in 2021, still yields fossils of the animal species consumed, the cutting marks left on the bones, the raw material and the cutting technique, with the chopper still in place. The latter was obtained using bipolar percussion, that is using an anvil and a hammer made of quartzite (Cauche *et al.* 2019). More than 200 pieces of lithic industry have been collected from the fossiliferous outcrops that are being eroded and limited to about 40 m in thickness, with a total sedimentary power of 170 m. The geomorphological conditions and the speed of annual erosion mean that each year new choppers and shards are collected from these silts. The construction is homogeneous and does not correspond to the African techno-typological tradition of the Oldowayen (Cauche *et al.* ibid.).

The few reactions questioning the discovery attest to a lack of knowledge of the geology, paleontology and geochronology research carried out by Indian researchers since the 1960s. The split quartzite pebble with fresh edges, collected *in situ*, was located in silts, i.e. the hydrodynamic context of low energy. It could not be naturally fractured by impacting blocks, the hundreds of thousands of pebbles exhumed from the silts that we have observed in 10 years of prospecting were deposited among the silts during flood recession and were rolled. The energy needed to break the quartzite is extremely powerful and did not exist in the plain during flood recession. Moreover, if this pebble had been fractured hundreds of kilometers upstream in the high Himalayan valleys, the edges would not be fresh but rolled. The only energy capable of moving this fractured pebble with two opposite impacts, found a few meters away from the traces of butchery, far from the other pebbles and isolated in the silts, is human force.

Figure 7.12. *Successive and superposed percussion marks of a sharp edge made of quartzite with a cortical bone removal on a 2.8 Ma bovid metapod, the intention was the consumption of marrow (photo: A. Dambricourt Malassé). For a color version of this figure, see www.iste.co.uk/dambricourt/embryogeny.zip*

Who were the Masol Hominids? Probably the same ones that lived some 4,000 km further east at Longguppo, in southern China, dating back 2.5 Ma (Han *et al.* 2017) where lithic tools of the same techno-typological tradition are associated with fauna (Boëda and Hou 2011). A fragment of a mandible with two worn molars was associated with these remains. Chinese paleoanthropologists recognize it as the oldest human fossil in China and thus in Asia. Other anthropologists prefer to see it as an "enigmatic" great ape. The doubt could be removed by microtomography of the molars to check the morphology of the contact between the dentin and the enamel (enamel-dentine junction or EDJ), as it corresponds to that of the cusps destroyed by wear. If these remains are not Hominids, the question remains as to how long Hominids have inhabited South Asia. This was probably the genus *Homo*. This thesis being the one with the most consensus, we should therefore expect to find lithic assemblages and human fossils on the Arabian plate from at least 3 million years ago, a perspective that is still inconceivable at the beginning of the 21st century and which postponed the emergence of the genus *Homo* more than 3 Ma.

7.7. Gracilization, cerebellization, anticipation and emotion

The gracilization of the Sapiens skull has been observed during the Holocene, i.e. for the past 10,000 years, the pneumatization of the frontal sinuses has disappeared, as well as the end-to-end occlusion visible until the Mesolithic on the Atlantic coast (Téviec, France), North Africa (e.g. Afalou) and Asia (e.g. Tam Hang in Laos) (Figure 7.13). It changed to psalidodontia, or cross bite occlusion, with the retraction of the lower alveolar arch under the maxillary arch. Anthropologists link this general gracilization to the change in lifestyle with the transition from a hunter-gatherer economy to that of agriculture and livestock.

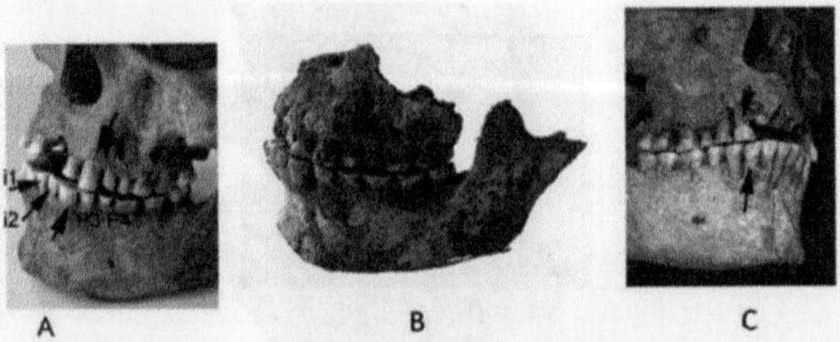

Figure 7.13. *End-to-end occlusion of incisors and canines, A: Epipaleolithic, Afalou (Algeria), B: Mesolithic, Plain of Jars, Tam Pong (Laos) and C: Teviec (Western Europe, Atlantic coast) (collections* Institut de Paléontologie Humaine, *photos: A. Dambricourt Malassé). For a color version of this figure, see www.iste.co.uk/dambricourt/embryogeny.zip*

With Sapiens, the contraction angles have reached their geometric and physical limits. And the reduction of the dental formula seems to be the future of occlusion, with the disappearance of the ends of the series (lateral incisor, second premolar and third molar). In other words, an evolution of the regulation of the neural crests would be in progress. This goes hand in hand with cerebellization, which also seems to shape the future of hominization. In 2005, the American anthropologist Anne Weaver published a study along these lines (Waever 2005), observing an inversion of the proportions between the brain and the cerebellum during the Holocene, thus, concomitantly with the gracilization and removal of the mandibular alveolar-dental arch under the maxillary arch:

> Cerebellar volume varies significantly relative to overall brain volume among mammalian orders, as well as within the primate order. There is also significant diachronic variation among fossil human taxa. In the australopithecines and early members of the genus Homo, the cerebral hemispheres were large in proportion to the cerebellum,

> compared with other hominoids. This trend continued in Middle and Late Pleistocene humans, including Neanderthals and Cro-Magnon 1, who have the largest cerebral hemispheres relative to cerebellum volume of any primates, including earlier and Holocene humans. In recent humans, however, the pattern is reversed; the cerebellum is larger with respect to the rest of the brain (and, conversely, the cerebral hemispheres are smaller with respect to the cerebellum) than in Late Pleistocene humans. The cerebellum and cerebral hemispheres appear to have evolved reciprocally. Cerebellar development in Holocene humans may have provided greater computational efficiency for coping with an increasingly complex cultural and conceptual environment. (Weaver 2005).

If the straightening reaches a physical limit, it would be consistent to observe not an angular evolution but a complexification, not only of the neural networks between the cerebellar and cerebral neocortex, but also of the vascularization of the cerebellum. Humanization can take over from physical straightening via cerebellization, to the benefit of emotion. Indeed, the recognition of the role of the cerebellum in so-called "higher" and emotional cognitive functions began to become clearer at the end of the 1990s with the research of Dr. Jeremy Schmahmann (1991), who describes the "Cerebellar cognitive affective syndrome" (CCAS):

> For several centuries, the cerebellum has been studied – and considered in clinical practice – solely for its involvement in motor coordination, ensuring, through cerebrocerebellar feedback circuits, fluidity and accuracy of movement. The motor symptoms occurring following cerebellar lesions, which are well known and well described, have long obscured their role in non-motor symptoms. It was not until the 1990s that case studies, and then, in the 2000s, more systematized research work on larger populations, made it possible to highlight the involvement of the cerebellum in the expression of so-called "higher" functions. The results of these studies coincided consistently with the discovery and description in neuro-anatomy of the existence of cerebello-cortical connections linking – albeit indirectly – the posterolateral and vermian regions of the cerebellum with non-motor cerebral areas, highly involved in cognition and affects. (Berthel-Tàtray 2017, author's translation)

Paleoanthropology has long supported the idea according to which the operational chains to produce stone tools were an invention of the first brain worthy of the name *Homo*, and that this capacity should correspond to a cerebral volume threshold or Rubicon. This concept is at the origin of the taxon *Homo habilis*. The innovations sought by anthropologists thus refer to the faculties of abstraction, those in particular that mentalize an operational chain, a sequence of effects that become the cause of the next step with a precise goal, for the production of a cutting tool. For the psyche, the sensitive experience is that of duration, the arrow of time, which does not conceive disruptions, discontinuities. *This is a major obstacle with which paleoanthropology has to contend because being discontinuous is counter-intuitive, whereas the discontinuity with the emergence of a reorganization of the pre-existing is the signature of the hominization process.*

7.8. The new Rubicon: brain stem verticality and cerebellar instability

7.8.1. *The body axis and Earth's gravity field*

A chimpanzee is not a Hominid. And the Rubicon is objective, the limit is physical and physiological in close relation to the Earth's gravity field. The concept that we have been supporting for several years offers more objectivity than the size of the brain hemispheres and the differences between their lobes, since the brain has continued to evolve between fossil and present-day great ape species. The discontinuity is physical, physiological and neurological first in the order of ontogeny and the angle of straightening of the basi-sphenoid, followed by the basi-occipital forming the clivus. This Rubicon is the threshold of the verticality of the brain stem and spinal cord, with numerous implications for the cerebellum, which controls balance.

The threshold of neuronal complexity is therefore objective and cannot be seen as an addition, but instead as an integration that requires the emergence of a new neuronal and vascular metasystem, to allow the neurons of the cerebellum to be informed about the balance of their own cerebellar fossa. *The cerebellum must be integrated into the loop of equilibrium that it controls and therefore a new neuronal loop is needed between the cerebral and cerebellar neocortices.*

A chimpanzee on the ground does not fall over, it can still maintain its balance walking on all fours. The Hominid, on the other hand, had to avoid falling over, the cerebellum had to anticipate more movements to avoid falling. The threshold of axial verticality is thus the emergence of a new reflexive loop of information more complex than that of any current and fossil species of great ape. This loop, or neural network, required the creation of new connections between the cerebellum and cortical areas of the brain specialized in "higher" cognitive functions. Cerebro-cerebellar networks are formed with the prefrontal (e.g. Watson *et al.* 2014; Parker 2016) and parietal cortex in language functions:

> These findings suggest that the right cerebellar lobule VIIb interacts with the posterior parietal cortex, specifically during the late stages of verbal encoding, when verbal information enters phonological storage. (Macher *et al.* 2014)

Cognitive functions are thus found to be connected to the anticipatory capacity of the cerebellum, and conversely, the cerebellum is linked to cerebral abstraction capacities, language faculties, memory and motor and non-motor learning, to complex visuo-spatial tasks with the mental construction of images and their rotation, to complex executive functions, attention, personality and emotion.

> The cerebellum is involved in planning, cognitive flexibility, abstract reasoning, working memory, verbal fluency and inhibition. Some studies suggest that the cerebellum can be activated during decision making or during coordination of two tasks at the same time, increasing velocity and automation of new movements. (Delecraz 2017, author's translation)

The future of paleocognition research is the integration of the cerebellum, via the brainstem, in the higher functions of the cerebral neo-cortex. These are the research topics of the ECPEC (Evolution of the Cerebellum, Posture, Emotion and Cognition) and OsteoCap (Osteology, Sensors and Paleoneurology) programs that we are developing in collaboration with the American Hospital of Paris and the University of Technology of Compiègne. Interactions with physiological factors that depend on the degree of verticality and the resulting tensions will be taken into consideration.

7.8.2. *Tools or utensils?*

It thus becomes opportune and interesting to distinguish the utensil from the tool.

Seizing a stone to defend oneself against a predator, or a stick to frighten a leopard, is an anticipatory gesture that does not transform the object into a weapon. A twig to catch termites, leaves to build a nest, becomes utensils but the object retains its integrity, it is the thoughtful choice that makes it a utensil. Using two stones together to break a nut, as chimpanzees (Carvalho *et al.* 2008) or capuchins (*Sapajus libidinosus*) do, means using two objects that become utensils because of their integration in a planned intention, that of consuming the contents of the broken object.

The difference between two stones used to crack a nut and the same two stones used to crack a third one, in order to produce cutting stones, is not an addition of utensils. The cognitive event is a reconceptualization of the use of stone, the emergence of the manufactured tool. The third stone is not an addition, it is no longer a utensil, but a project in the psyche and it takes on a completely different meaning. This meaning is the conceptualized finality in which the three stones form a whole, or the indispensable parts of a whole; it is what André Leroi-Gourhan described as an "operational chain (*chaîne opératoire*)".

The cerebellum is not added to the brain, the process is a reorganization and complexification of neural networks with very probably more neuroblastic divisions to form the cerebral and cerebellar neocortex, as well as an integration of the latter in prefrontal and parietal activities. What we posit as an inference is a link of continuity between the emergence of reconceptualization by integration of a useful object, and the emergence of the suspended, and therefore more unstable, cerebellum. The latter is necessarily integrated as an indispensable organ in a new neuronal loop of information reflection and which associates it with the cognitive properties of the brain. The integration of the cerebellum into a reorganized reflexive loop is reflected in the reconceptualization of the use of the stone.

When a Hominid fractures a pebble with two other stones – the anvil and the hammer – either to extract shards or to sharpen it on an edge, the gaze accompanies the gesture guided by a reflection that knows the sequence of movements, but is also aware of the hands and fingers. This body consciousness was built with the image of the body when the nervous system was in its psychomotor learning phase. The chimpanzee who breaks a nut while holding the stone with one hand is certainly aware of its hands, or, in other words, it knows that it has two hands. But does the chimpanzee reflect on the intellectual mechanism? Just as the emergence of an operational chain for the production of lithic tools is not the addition of a third stone, so the new modalities of the psychomotor development of the Hominian posture are not an addition of neurons.

The spatial organization of the cutting marks on Masol fossils, the oldest ones attributable to *Homo*, not only attests to a technical intelligence but also to an anatomical knowledge of the animals consumed. The cutting marks are on zones of aponeurosis insertion which maintain the articulation of the bones; they are precise, fast, the gesture of rotation of the wrist is inscribed in the trajectory of the groove; incisions very close together over 1 cm long make it possible to think that the Hominid held a splinter between thumb and index finger. *At 2.8 million years old, these are the oldest testimonies of a very precise knowledge of the anatomy of consumed animals and of a plan for their dismemberment* (Dambricourt and Cauche 2022).

7.8.3. *From the reflection of the image to the symbolic creation of its meaning*

Australopithecus and *Homo* are not augmented great apes, and a chopper is not a third or fourth pebble added to a possible operational chain of the nutcracking great ape fossil species. The biological continuity is phylogenetic, and it is quite plausible that techniques such as fracturing nuts with an anvil and hammer have been passed down in several great ape clans in the midst of mutagenesis. But the threshold of neural complexification is a reality that is (1) physical, (2) physiological and (3) psychological. It therefore corresponds to the emergence of a nervous system that is verticalized at birth and a reflexive loop of information that necessarily increases its level of complexity in order to allow the cerebellum to control the more unstable posture of the cerebellar fossa.

There follows a whole chain of causalities with psychomotor development, of which the invention of operational chains is only one aspect. The image of the body is constructed in relation to the parent and the emotions during psychomotor development; it is accompanied by an awareness of existing as a distinct and singular individual, with an emotional charge that does not have an equivalent in the great apes, which is the fear of falling. This reflection of the image of the body becomes the cornerstone of the process of humanization to be distinguished from the process of hominization of the body. This distinction is fundamental.

Hominization is deterministic, this is the phylogenetic part of the process, but humanization is plastic and non-reproducible without the post-natal transmission of the experiences acquired over the course of generations, on the one hand, and without this psycho-affective relationship between the nursing mother and her newborn, that is without the learning phase, on the other hand. The hereditary memory of the process is transmitted by the gametes, but the memory of the experiences and knowledge that begin with this conscious relationship between the mother and her newborn are not. The identification of the self and the emotional construction of its image are thus impossible without the otherness, without the opening of the mother to this unknown which is her newborn, without the risk-taking and courage of the mother for the protection of her unknown little one.

The Hominids would have developed capacities of information analysis longer than that of a great ape, and this induction is not an anthropocentric hypothesis, but the conclusion of a broad set of converging data that we have synthesized over 40 years of research.

Humanization has raised the level of reflection over the centuries and millennia, which in turn has raised anthropological questions, that is to say, identity or ontology: Is the being debated by philosophers an illusion of neural networks? Who is speaking or what is speaking? Who is saying "I" or what is saying "I"? How should we express this learning phase: should we talk about the conscious body, while self-awareness is located in the brain? Should we talk about body consciousness, but what is body consciousness in terms of nerve impulses? Could the sense of the sacred and of sacrilege be just a neuronal invention devoid of rationality? Could the being's sense of integrity just be an illusion? Undermining the integrity, the dignity of the being, would this be an illusion of physico-chemical processes which have no dignity as such? The natural philosophers developed the basic discernment between physics, which is only the setting of an accessible part of reality into the form of equations, and the reality which contains Man and his equations, or the physical sciences. That the real has a level of reality accessible to symbolic thought – a qualitative world – seems to be more relevant than the physicalist doctrine – or the materialist doctrine which reduces the real to physical theory – and which does not address the ontological implications of the recognition of the crime against Humanity. It would be to admit Being beyond the curvature of space–time, Being beyond one's own neural networks, a "presence of mind" that itself allows metaphysical questioning thanks to its freedom of thought (see Volume 1).

Self-awareness integrates not only the image of the body, but also emotions, experiences, dreams, fantasies and memories, and this image does not have a physical form. It is invisible and personal to each individual.

What is the image of this invisible personality? Is it a form of symbolic thought? And is it not on the basis of this symbolic image that communication could convey meaning, or understanding? This self-awareness can be identified by a name or a vocalization and thus recognized and signified by others, it makes sense and gives meaning to all individuals. Conscious and thoughtful communication thus appears necessarily symbolic, such as the representation of other members of the clan, other organisms, whether animal or vegetable.

The symbol is thus charged with meaning, it is the expression of the invisible thought.

Plato was therefore not wrong to consider that meaning is "hidden" from view and accessible only by the psyche. What the philosophers considered as an archetype would be a symbol developed in the human psyche carrying meaning, whereas the world, understood from the outside by measurements, or physis, does not give access to the meaning, or to the symbol, but access to these laws, to its rules or its principles. The human psyche is much more than a mathematical measure, Leonardo da Vinci understood this (see Volume 1). Measurement provides access to the physical laws, but the meaning or the symbol is beyond the form and the movement that animates it. And it is the artist or the shaman who gives it form by creating it with matter, and using objects such as feathers, teeth, colored stones and bones (Dambricourt Malassé 2021b).

The first evidence of a consciousness challenged by its own physical image, or that of a deceased fellow (a skeletal face), is a 3-million-year-old pebble that was found in 1925 in Makapansgat. The pebble was a manuport because the karst was found far from possible lithic raw materials (Bednarik 1998). A face is recognizable with a ring above two eyes, a nose and a mouth with a row of teeth, which are all the effects of erosion. It is therefore not a question of a work of art, but a testimony to self-awareness that was already particularly evident for having been challenged by a stone figure. Whether the Hominid was *Australopithecus* or *Homo*, the debate remains open, but it does not matter, both had crossed the threshold of verticality and corresponded to new neuronal and hormonal organizations, more complex in their capacities for analysis and synthesis, for reflection and innovation, for thoughtful awareness of their body, for themselves, and for the outside world, both the living and deceased.

The image of the body and the self-image reinforced the founding language of otherness, certainly present in the great apes, but verticalization modified it radically. The cerebellum now had its say in some way, with emotion and precision. It is very likely that young children had to develop behaviors and reactions that were surprising because they were unknown by the codes transmitted by education. As a result, it was no longer only the genus *Homo* that had new faculties of reflection, abstraction and creative and emotional innovations, but also *Australopithecus* and *Paranthropus*.

7.9. New sciences for an emerging evolutionary problem

The hominization process was thus manifested at thresholds that corresponded to a break in the metabolic equilibrium of the reproductive cells which resulted during embryogenesis in these angular jumps due to the contraction of the fourth ventricle followed by the straightening of the neural tube. These jumps are undoubtedly the result of the duration of migration of the neural crests and of modifications all along the embryo transmitted by the connective tissues, since the scale is in the order of a few centimeters.

And it is hardly possible to explain the iteration of the straightening amplification without a memory stored in the sex cells. The gametes therefore must have been endowed with reorganization properties sensitive to production thresholds of their own complexity.

The production of complexity is necessarily coupled with the memorization of the regulation of new states of equilibrium along the cranio-spinal axis of the embryo. These mechanisms should be placed on a geological time scale. Nothing happened for millions of years, and for the vast majority of lineages, nothing has changed to date. Only a few species have reached a threshold of instability in their gametogenesis. The other species have transmitted a memory in a stable state of equilibrium up to the current species. Moreover, these thresholds of instability have not followed a regular rhythm. The stages lasted more than 10 million years between the first small simian (monkey) and the first great apes, then between the latter and the first Hominids. This threshold is one of unprecedented instability with permanent verticality and processes that suddenly accelerated that led to the extinction of three fundamental ontogenies, which had never happened before: *Australopithecus*, *Paranthropus* and *Homo* disappeared. The only fundamental ontogenesis that remains is Sapiens, us, here and now, with an exponential population density and technical intelligence that has reached the solar system and enabled the conquest of space.

These are processes such as Ilya Prigogine (1917–2003) defined with dissipative structures and for which he received the Nobel Prize in Chemistry in 1977 (Prigogine and Stengers 1979, see Chapter 1, Volume 1). In order to become more complex, the system must be far from equilibrium. Gamete lineages have thus remained far from equilibrium by amplifying the instability of their organization. At threshold limits, the state of equilibrium has bifurcated, creating a new level of organization that guarantees stability. These bifurcation phenomena are modeled by so-called "chaotic" attractors because the new state is an innovation that nothing prefigures, it is determined by the context – sensitivity to initial conditions – but the reorganization is totally innovative. However, there is a considerable difference with the phylogenetic curve of the straightening. At the bifurcation threshold, the innovation was not indeterminate, because an iteration of an internal logic occurred with predictable cascading effects resulting from longer cell divisions. For example, the five cusps of the dm1 of the *Hominidae* are an innovation, but they were inevitable since the pentacuspid shape is logical for molar formation. dm1 could form five cusps like all other molars if they had enough time. And the most predictable consequence was obviously the threshold of verticality of the axial endoskeleton.

The mathematical model will therefore not be that of chaotic attractors, which is why the term harmonic attractor has been proposed in reference to conservation of the memory of processes (Dambricourt Malassé 1992, 1995, 1996), whereas their dissipation would have generated a non-reproducible transformation in its effects and would have been modeled by chaotic attractors. This neologism was validated at the CNRS in 1992 by the mathematician René Thom (1923–2002) – who received the Fields Medal in 1958 – considering that it is the systems that control their morphogenesis and not external unpredictable agents. The "New Alliance" of Ilya Prigogine restoring human consciousness to its place in the internal history of the cosmos, following Pierre Teilhard de Chardin, was made possible by carbon chemistry (organic chemistry) (Brack and Raulin 1991), but René Thom's mathematics are best able to account for the stability of the phylogenetic trajectory of the straightening beyond the succession of thresholds of increasing instability. René Thom was a convinced evolutionist and his catastrophe theory had nothing to do with the catastrophism and creationism of Georges Cuvier (see Volume 1). The art of this research is in reconciling form and duration (irreversibility). Thom showed that there are only seven of them: the fold, the cusp, the swallowtail, the butterfly and three different forms of umbilicals. This is the field of possibilities allowed by mathematics for the evolution of species, and so it has its internal logic and the potential for innovation through combinations of these forms and the laws of physics.

To conclude, self-referencing processes would be modeled by harmonic attractors. Human paleontology would therefore benefit from integrating the science of complex dynamic systems to become part of the sciences of the 21st century.

7.10. The future of Sapiens, a dialog between the cerebellum and the brain?

The study of the mandible of primates followed the phylogenetic classification, which corresponds to the closure of the sphenoidal angle, as recalled by Delattre and Fenart (Figure 7.14). Some morphogenetic trajectories required downgrading, such as the mandible of the Chadian anthropoid *Sahelanthropus* – a great ape, not a Hominid – in accordance with a rectification of the virtual reconstruction of the position of the petrous pyramids and the occipital hole. Similarly, the skull base of *Ardipithecus* is that of a more recent *Australopithecus* and must be removed from the composite reconstruction, since its occluso-postural balance matches the degree of straightening of the great apes. The development of the Neanderthals was never that of *Homo sapiens* and neither was the 300,000-year-old skull from Jebel Irhoud in North Africa. In the current state of paleontological knowledge, the oldest *Homo sapiens* was discovered in Herto in Ethiopia and dates back 160,000 years. These few fossils naturally find their place in the curve of the evolutionary process at the origin of neural straightening.

This process is phylogenetic, that is it was transmitted during gametogenesis and immediately acted on the regulation of embryogenesis according to the cerebrospinal gradient, generating increasing complexity channeled to the nervous system (Figure 7.14).

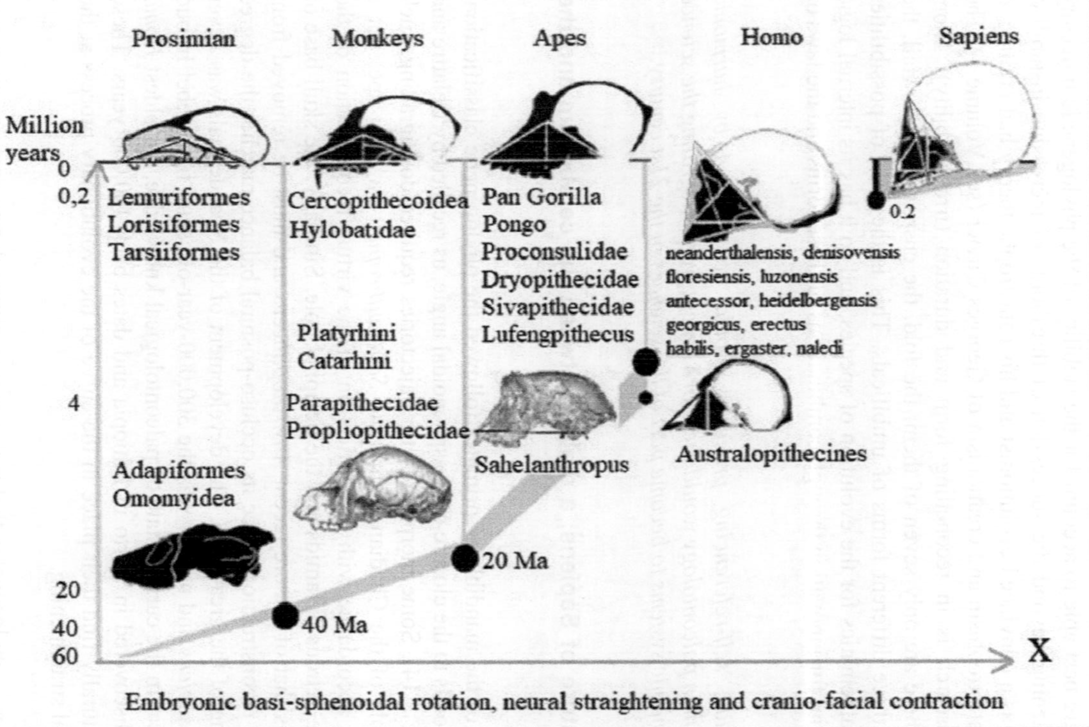

Figure 7.14. *The phylogenesis of successive fundamental ontogeneses illustrated at the adult cephalic pole by rotation of the basisphenoid, the staggering of the three cranial fossae and the cranio-facial contraction (A. Dambricourt Malassé 1993, 1996; Dambricourt* et al. *1999; Chaline* et al. *2000)*

The phylogenesis of *Homo sapiens* thus corresponds to the amplification of the closure of the fourth ventricle of the neural tube and the straightening of the dorsal cord. These embryonic dynamics were followed by complexifications of neurogenesis with prolongations of cerebral and cerebellar neocorticalizations and their connections through the brainstem. The increasing complexity of the cerebral hemispheres thus appears to be the most visible part of a complexity generalized to the entire nervous system, since it was accompanied by cerebellization as the instability of the axial endoskeleton increased. We are the last threshold of this very long phylogenesis (Dambricourt Malassé 2017).

The introduction of embryogenesis has thus confirmed the theory of a process known as "hominization", i.e. the complexification of the central nervous system, which takes longer to reach maturity, with a delay in endocrine functions, facial growth and psychomotricity, with major cognitive implications at the threshold of verticality, for locomotion that has become exclusively bipedal. *The threshold which saw the emergence of the Hominidae is thus not the disappearance of arboricolism but the disappearance of quadrupedalism on the ground and the emergence of a verticality to be assumed with more emotion and apprehension, both with regard to gravity to avoid falling and to predators with regard to the mother–child pairing. Otherness has grown from a strengthening of the bonds between members of the same clan.* Will the brain-cerebellum dialog be the future of humanization? And would the continuation of intrauterine hominization presage a pre-adaptation to living without gravity?

2001: A Space Odyssey, a cinematographic masterpiece by Stanley Kubrick produced in 1968, ends with an unexpected vision, but one whose symbolism is extremely striking: the astronaut who has left the Earth and reached the confines of curved space–time, regenerated into a fetus with his eyes opening onto the blue planet. However, the terrestrial fetus does not need eyes. What is the significance of this reversal of the fetal gaze on the future of hominization whose umbilical cord is the cosmogenesis and the cosmos its placenta?

Conclusion

Irreversibility, Responsibility, Otherness

Since Antiquity, philosophers have been searching for the singularity that distinguishes human beings from other animals. This singularity is also their own, that is, the ability to reason or reflect consciously on the human body, on the world and its relationships with the world, making it intelligible in order to make oneself intelligible. It is necessary to give oneself the means to do this through reasoning and the objective method. The method aims at freeing oneself from subjectivity, belief and illusions, within the limits of the tools invented by human intelligence. The need to distinguish illusions from reality, reflection on the being and on shifting movement, has generated the distinction between nature, or *physis*, on the one hand, and what reflects on it, questions it and meditates on it, in other words, the human psyche. This distinction is always present in the minds of scientists and philosophers of nature. Paleoanthropology is thus at the crossroads of disciplines, since the subject of study is also the identity of *Homo sapiens*, seeking to understand itself, a living, thinking, conscious identity, which cannot be confused with a mathematical model. Through science, the human being reaches reliable knowledge, but the human being will always exceed its tools and its measures; its imaginary worlds, emotions, dreams and lightning flashes of inspiration also make a human a singular animal being.

The future of hominization and humanization that will emerge from our biological and cultural singularities is a real concern and that is why the understanding of these processes, which are our common and universal identity transmitted from a unique phylogenesis, cannot be the property of a scientific ideology and of any evolutionist dogma. We have seen how much this quest for understanding Man's place in nature and the universe has clashed, over the centuries, with scholasticism and its arguments of authority; the awareness of evolution, which is very recent, is a liberation, a threshold in the process of humanization and a freedom with all that it implies: exploration without any other guide than the scientific compass and reasoning. However, evolutionary anthropology has had to defend itself for decades against a new scholasticism, not the neo-Darwinian doctrine, but neo-Darwinism which is explicitly the opposite of orthogenesis for reasons clearly stated since the Paris colloquium in 1948, at the *Institut de Paléontologie Humaine*. This questioning is external to science, and the scientific theory of evolution, including hominization, has to take into consideration the scientific, metaphysical and artistic mind, closely linked to verticality, itself closely linked to the cosmogenetic process of increasing complexity-consciousness. The threshold of verticality that has emerged from this process has manifested as an awareness of one's own existence for the last 3 million years, since hominids first recognized their own image in a stone. The collection of the manuport from Makapansgat is the first known manifestation of a natural curiosity, which led to the creation of cabinets of natural curiosities, followed by museums of natural history and their museums of Man.

Research on the processes that generated the verticalized anatomy of *Homo sapiens*, or Anatomically Modern Human, is focused on a disappeared reality, all the more inaccessible as it is constructed from an internal arrow of time specific to gametogenesis. Science can then only restore geometric models from fossilized forms. To transform this data into scientific knowledge, it must be transcribed by equations that describe the changes in shape. The method we have developed for correlating angles is the guiding thread. It is an idealization of science, because until now, no equation has ever described the trajectory of craniofacial contraction in three dimensions, of any fossil and current primate species. But it shows that, theoretically, if all stages of development of all extinct and current species were known, a super-calculator could write an equation for each fundamental ontogenesis. The equation sought would then be the one that describes the evolution from one fundamental ontogenesis to another, the most complete scientific formalization of the trajectory of axial straightening and craniofacial contraction.

Until the 21st century, it was therefore barely possible to understand processes that have disappeared and were irreversibly self-sustaining, other than in a theoretical way through this approximated equation. The least debatable observation is the stability of the phylogenetic trajectory, the equation is therefore deterministic, that is, the time spent is a parameter, the memory remains and it was a deciding factor.

The equation will probably never be written, but thanks to the angular data and their correlations, this method makes it possible to conclude that there were evolutionary processes particular to gametes that were never observed before and were thus absent from the evolutionist doctrines of the 19th and 20th centuries. The most inadequate was Princeton's neo-Darwinian doctrine, which postulated accidental genetic mutations without causality, without history and without memory. No duplication during gametogenesis has deviated or definitively stopped the process of internal production of complexity and, especially, the management of these thresholds. Gametogenesis does not always have a clean slate; the duration is recorded as a memory of the mechanisms that control the unpredictable variations that put the fertilized oocytes in danger of extinction.

Stability and irreversibility thus make one wonder about the properties of gametes when they reproduced, because they are found today in the Sapiens *gametes. They are indeed properties of construction of an organism, like a multiform living being.*

Thus, the primitive organization, initially horizontal since the first *Chordata*, ended up aligning itself with the Earth's gravity field. It is from this increasing instability with a loop of reflection of information between the brain and the cerebellum, that *Sapiens*, the observer and the philosopher of nature, emerges *in fine*, vertically posed on the Earth's surface, in the continuity of the genesis of particles, the first cells and the complexification of the nervous system of organisms. The axis of the world for Aristotle, the thinking arrow of evolution for Pierre Teilhard de Chardin, *Homo sapiens* is indeed a bit of both, confused with the Earth's gravity field in the continuum of a growth axis of internal complexity of the universe. This has reached the stage of life on our planet, or the transmissible memorization of these complex processes. When Teilhard wrote that hominization is, first of all, a cosmic phenomenon, the 21st century can only note this without sinking into obsolete anthropocentrism. On the contrary, this is a threshold of maturity, of consciousness and an awakening of the sense of responsibility.

A Discovery Channel documentary film directed by the Thomas Johnson aired this discovery in 2005 (Johnson 2005), integrating craniofacial biodynamics by Marie-Josèphe Deshayes, heterochronies of development by Jean Chaline and Didier Marchand (1945–2016), homeotic genes by Denis Deboule, commentaries by Phillip Tobias (South Africa), Wu Xin Zhi (I.V.P, Beijing) and Yves Coppens (France). This same documentary integrates the historical debate that took place during the international congress of the *Institut de France* at the Academy of Sciences in 2004, with Yves Coppens, Michel Brunet, Brigitte Senut, Timothy White and during which this discovery was developed (Caro 2004; Dambricourt Malassé 2006).

The philosopher of human sciences, Edgar Morin, integrated the discovery of harmonic attractors as early as 2001 (Morin 2001).

We are the last threshold of straightening and the first conscious stage of a terrestrial process with cosmic origins that is still in progress. We are the first threshold of cerebro-cerebellar complexity that allows us to consciously reflect on the vanished past from which we emerge, to question the beginning of all that has been observed, since in the beginning lies, it is believed, the explanation of the presence of an organism, its permanence and reproduction. We have this capacity to anticipate by mentally, or thanks to super-calculators, testing the consequences of our actions and choices.

Between irreversibility and responsibility, there is still time to take in hand the effects of this acceleration in the production of complexity, because it is not impossible that public health problems related to disorders of the occluso-postural balance will manifest a new oscillation in the regulation of embryogenesis. The French association *Fédération des Recherches sur l'Évolution Humaine, l'Ostéopathie et la Posture au service de la Santé* (FREHOPS, Federation of Research on Human Evolution, Osteopathy and Posture in the service of Health)[1] was created in 2018 with this in mind. The initiative is part of this responsible awareness of knowing that we are part of an ongoing process without pretending to know its ultimate outcome other than to keep ourselves alive, here and now, by listening to emerging infantile differences and by understanding the planet as the placenta of all evolutionary processes. We are not the final summit of a creation but we have the means to reflect on our responsibility over a long duration toward our verticalized fundamental ontogenesis and its bonds with plant and animal species in an extended otherness.

1 Available at: www.frehops.fr.

References

Ahern, J.C.M. (1998). Underestimating intraspecific variation: The problem with excluding Sts 19 from *Australopithecus africanus*. *Am. J. Phys. Anthropol.*, 105, 461–480.

Arambourg, C. (1950). Commentaires. In *Paléontologie et transformisme*, Arambourg, C., Cuénot, L., Grassé, P.-P. (eds). Albin Michel, Paris.

Arambourg, C. and Mouta, F. (1952). Les grottes et fentes à ossements du Sud de l'Angola. *Actes du Congrès panafricain de préhistoire*, 12, 301–304.

Barboni, D., Ashley, G.M., Bourel, B., Arráiz, H., Mazur, J.C. (2019). Springs, palm groves, and the record of early hominins in Africa. *Review of Palaeobotany and Palynology, Special Issue: Paleofloras, Paleovegetation and Human Evolution*, Carrión, 266, 23–41.

Been, E., Gómez-Olivencia, A., Kramer, P.A., Barash, A. (2017). 3D reconstruction of spinal posture of the Kebara 2 Neanderthal. In *Human Paleontology and Prehistory Contributions in Honor of Yoel Rak*, Marom, A. and Hovers, E. (eds). Springer, Cham.

Benoit-Levy, R. (n.d.). Théories plus récentes [Online]. Available at: https://www.osteopatheaparis.fr/specialites-2/orthodontie/croissance-cranienne/theories-plus recentes/.

Bergamini, M., Pierleoni, F., Gizdulich, A., Bergamini C. (2008). Dental occlusion and body posture: A surface EMG study. *Cranio: The Journal of Craniomandibular Practice*, 26(1), 25–32.

Berthel-Tàtray, M.C. (2017). Le syndrome cognitivo-affectif du cervelet. *Revue Neurologique*, 173, Supplement 2, S33–S208.

Bhandari, A., Kay, R.F., Williams, B.A., Tiwari, B.N., Bajpai, S., Hieronymus, T. (2018). First record of the Miocene hominoid Sivapithecus from Kutch, Gujarat state, western India. *PLoS One*, 13(11), e0206314.

Black, D. (1925). Asia and the dispersal of primates. A study in ancient geography in Asia and its bearing on the ancestry of man. *Bulletin of the Geological Society of China*, 4(2), 133–183.

Black, D. (1926). Tertiary man in Asia: The Choukoutien discovery. *Nature*, 118, 733–734.

Black, D. (1927). The lower molar hominid tooth from Chou Kou Tien deposit. *Paleontologia Sinica*, Series D, 7(1), 1–29.

Black, D. (1930). On an adolescent skull of *Sinanthropus pekinensis* in comparison with an adult skull of the same species and with other hominid skulls, recent and fossil. *Paleontologia Sinica*, Series D, 7(2), 1–114.

Black, D. (1933). On the endocranial cast of the adolescent Sinanthropus skull. *Proc. Roy. Soc. London (B)*, 112, 263–276.

Boëda, E. and Hou, Y.M. (2011). Le site de Longgupo. Chongqing-Chine. *L'Anthropologie*, 115, 1.

Bois, E. (2010). La dynamique posturale non-linéaire. Partie I : Bases et fondements. *L'Anthropologie*, 114(3), 370–387.

Boule, M. (1911–1913). L'homme fossile de la Chapelle-aux-Saints. *Annales de paléontologie, Masson*, Paris.

Brodland, G.W., Conte, V., Cranston, P.G., Veldhuis, J., Narasimhan, S., Hutson, M.S., Jacinto, A., Ulrich, F., Baum, B., Miodownik, M. (2010). Video force microscopy reveals the mechanics of ventral furrow invagination in Drosophila. *Proc. Natl. Acad. Sci. USA*, 107(51), 22111–22116.

Broom, R., Robinson, J.T., Schepers, G.W.H. (1950). Sterkfontein ape-man Plesianthropus. *Transvaal Museum Memoir*, 4, 104.

Brunet, M., Guy, F., Pilbeam, D., Taisso Mackaye, H., Likius, A., Ahounta, D., Beauvilain, A., Blondel, C., Bocherens, H., Boisserie, J.-R. *et al.* (2002). A new hominid from the Upper Miocene of Chad, Central Africa. *Nature*, 418, 145–151.

Brunet, M., Guy, F., Pilbeam, D., Lieberman D.E., Likius, A., Mackaye, H.T., Ponce de León, M.S., Zollikofer, C.P.E., Vignaud, P. (2005). New material of the earliest hominid from the Upper Miocene of Chad. *Nature*, 434, 752–755.

Callaway, E. (2018). Controversial femur could belong to ancient human relative. *Nature*, 553, 391–392.

Caro, P. (2004). Entretien avec Anne Dambricourt Malassé. Développement embryonnaire et évolution. *La Lettre de l'Académie des Sciences*, 13, fall 2004.

Cauche, D., Singh, M., Dambricourt Malassé A., Tudryn, A., Gargani, J., Abdessadok, S., Chapon Sao, C., Moigne A.M., Karir, B., Pal, S. (2019). The first chopper in Asian late Pliocene silts, Masol, Northwestern India: Bio-lithostratigraphy, magnetostratigraphy and dating evidences. *The Journal of the International Union of Prehistoric and Protohistoric Sciences* [Online]. Available at: https://hal.archives-ouvertes.fr/hal-02374833.

Cerling, T.E., Mbua, E., Kirera, M., Kyalo Manthi, F., Grine, F.E., Leakey, M.G., Sponheimer, M., Uno, K.T. (2011). Diet of Paranthropus boisei in the early Pleistocene of East Africa (supporting information). *Proc. Natl. Acad. Sci. USA*, 108(23), 9337–9341 [Online]. Available at: https://www.ncbi.nlm.nih.gov/pmc/articles/PMC3111323/.

Chaline, J., Dambricourt Malassé, A., David, B., Magniez-Jannin, F., Marchand, D., Courant, F., Millet, J.J. (1996). Quantification de l'évolution morphologique du crâne des Hominidés et hétérochronies. *Comptes Rendus de l'Académie des Sciences de la Terre et des Planètes*, 326, 291–298.

Chaline, J., Durand, A., Marchand, D., Dambricourt Malassé, A., Deshayes, M.J. (1998). Chromosomes and the origins of apes and australopithecines. *Human Evolution*, 11, 43–60.

Chaline, J., Durand, A., Dambricourt Malassé, A., David, B., Magniez-Jannin, F., Marchand, D. (2000). Were climatic changes a driving force in hominid evolution? In *Climates, Past and Present*, Hart M.B. (ed.). Geological Society, London.

Chamyal, L., Dambricourt Malassé, A., Maurya, D.M., Raj, R., Bhandari, S., Pant, S.K., Gaillard, C. (2011). Discovery of a robust *Homo sapiens* in India (Orsang, Lower Narmada Basin, Gujrat). Possible continuity with Indian Homo erectus. *Acta Anthropologica Sinica*, 2, 167–191.

Chapon Sao, C., Abdessadok, S., Dambricourt Malassé, A., Singh, M., Karir, B., Bhardwaj, V., Pal, S., Gaillard, C., Moigne, A.M., Gargani, J., Tudryn, A. (2016). Magnetic polarity of Masol 1 locality deposits, Siwalik Frontal Range, Northwestern India. *Comptes Rendus Palevol, Special Issue: Human Origins in the Indian Sub-continent*, 15(3–4), 407–416.

Chapon Sao, C., Tudryn, A., Dambricourt Malassé, A., Moigne, A-M., Gargani, J., Singh, M., Abdessadok, S., Voinchet, P., Cauche, D., Karir, B., Pal, S. (2018). The late Pliocene prehistoric site of Masol (Northwestern India): New dating of the geological and paleontological context of cut marks and stone tools. *XVIIIème Congrès de l'Union Internationale des Sciences Préhistoriques et Protohistoriques (UISPP)* [Online]. Available at: https://uispp2018.sciencesconf.org/175540.

Chen, X. and Brodland, G.W. (2008). Multi-scale finite element modeling allows the mechanics of amphibian neurulation to be elucidated. *Phys. Biol.*, 5(1), 015003.

Cho, K.H. (2005). Asymétries cranio-faciales et dysmorphoses : analyse architecturale cranio-faciale des populations médiévales de La Queue-en-Brie (Val-de-Marne, Île-de-France). PhD Thesis, Muséum national d'Histoire naturelle, Paris.

Clarke, R.J. (1977). The cranium of the Swartkrans Hominid SK 847 and its relevance to human origins. PhD Thesis, University of the Witwatersrand, Johannesburg, South Africa.

Clarke, R.J. and Kuman, K. (2019). The skull of StW 573, a 3.67 Ma *Australopithecus* skeleton from Sterkfontein Caves, South Africa. *J. Hum. Evol.*, 134, 102634.

Coppens, Y. (1975). Évolution des Hominidés et de leur environnement au cours du Plio-Pléistocène dans la basse vallée de l'Omo en Éthiopie. *Comptes Rendus de l'Académie des Sciences*, 281, 1693–1696.

Coppens, Y. (1980). The differences between *Australopithecus* and Homo: Preliminary conclusion from the Omo Research Expedition's studies. In *Current Argument on Early Man, Report from a Nobel Symposium*, Königsson, L.-K. (ed.). Pergamon Press, Oxford.

Cunningham, D.L. and Jantz, R.L. (2003). The morphometric relationship of Upper Cave 101 and 103 to modern *Homo sapiens*. *J. Hum. Evol.*, 45(1), 1–18.

Curnoe, D. (2010). A review of early Homo in southern Africa focusing on cranial, mandibular and dental remains, with the description of a new species (*Homo gautengensis* sp. nov.). *Journal of Comparative Human Biology*, 61, 151–177.

Curnoe, D. and Tobias, P.V. (2006). Description, new reconstruction, comparative anatomy, and classification of the Sterkfontein Stw 53 cranium, with discussions about the taxonomy of other southern African early Homo remains. *J. Hum. Evol.*, 50, 36–77.

Dambricourt Malassé, A. (1987). Ontogenèses, paléontogenèses et phylogenèse du corps mandibulaire catarhinien. Nouvelle interprétation de la mécanique humanisante (théorie de la fœtalisation, Louis Bolk, 1926). PhD Thesis, Muséum national d'Histoire naturelle, Paris.

Dambricourt Malassé, A. (1988). Hominisation et fœtalisation. *Comptes Rendus de l'Académie des Sciences*, 307(2), 199–204.

Dambricourt Malassé, A. (1989). Etude de la base crânienne d'un fœtus humain monté sur lame. Report, Fondation Fyssen, Paris.

Dambricourt Malassé, A. (1992). L'hominisation et la théorie des systèmes dynamiques non-linéaires (Chaos). *Revue de Biomathématique*, 117, 5–32; 118, 5–48; 119, 5–54.

Dambricourt Malassé, A. (1993). Continuity and discontinuity during modalities of hominization. *Quaternary International, Special Issue: Modes and Tempos in Evolution in the Quaternary*, 19, 85–100.

Dambricourt Malassé, A. (1995). Les attracteurs inédits de l'hominisation. Ontogenèse fondamentale, attracteurs chaotiques et attracteurs harmoniques. *Acta Biotheoretica, Société Francophone de Biologie Théorique*, 43, 113–125.

Dambricourt Malassé, A. (1996a). Exemple du chaos appliqué en paléontologie humaine. In *Chaos et fractales dans l'activité solaire*, Rozelot, J.P. (ed.). École thématique du CNRS, Paris.

Dambricourt Malassé, A. (1996b). Nouveau regard sur l'origine de l'Homme. *La Recherche*, 286, 46–54.

Dambricourt Malassé, A. (1998). L'hominisation de la sphère basi-cranio-faciale : un processus d'origine embryonnaire, en accélération continue et non-chaotique. Implication pour l'orthopédie dento-maxillo-faciale. *Autredent*, 21, 5–10.

Dambricourt Malassé, A. (1999). De la morphogenèse cranio-faciale embryonnaire à l'architecture adulte. Problèmes et méthodes pour la reconstitution des ontogenèses fossiles. *Mémoires de l'Académie des Sciences, des Art et Belles-Lettres de Caen*, 37, 7–74.

Dambricourt Malassé, A. (2000). Embryogenèse et évolution humaine. *Les Cahiers du Centre d'Étude de l'Ostéopathie, de la Posturologie et Sciences Annexes*, 1, 2–4.

Dambricourt Malassé, A. (2001). Nouvelles approches de l'évolution crânienne des *Homo erectus* de Java. In *Origine des peuplements et chronologie des cultures paléolithiques dans le Sud-Est asiatique*, Sémah, F., Falguères, C., Grimaud Hervé, D., Sémah, A.M. (eds). Artcom, Paris.

Dambricourt Malassé, A. (2005). Un nouvel angle de la base pour une meilleure compréhension de la morphogenèse basi-crânienne des Hominidés et des modalités évolutives. In *Biométrie et anthropologie de la tête, de la face et du cou. Biométrie Humaine et Anthropologie*, 23 (1–2), 53–62.

Dambricourt Malassé, A. (2006a). Évolution du chondrocrâne et de la face des grands anthropoïdes miocènes jusqu'à *Homo sapiens*, continuités et discontinuités. *Comptes Rendus Palevol*, 5, 109–117.

Dambricourt Malassé, A. (2006b). La question des origines de la lignée humaine, les modèles évolutionnistes confrontés aux fossiles, implications en orthopédie et posturologie. *Les Cahiers du Centre d'Étude de l'Ostéopathie de la Posturologie et Sciences Annexes*, 4, 38–57.

Dambricourt Malassé, A. (2009). Embryogeny and human phylogeny. In *Asian Perspectives on Human Evolution*, Sankhyan, A.R. (ed.). Serial Publications, New Delhi.

Dambricourt Malassé, A. (2010). The relationships between occlusion and posture in the hominid lineage, implications for the transition between Mesolithic and Neolithic populations. *International Journal of Modern Anthropology*, 3, 14–63.

Dambricourt Malassé, A. (2011a). The human lineage, a macro-evolutionary process acting during embryogeny with emergent macro-evolutionary implications. In *Biological Evolution: Facts and Theories*, Auletta, G., Leclerc, M., Martinez R.A. (eds). Gregorian & Biblical Press, Rome.

Dambricourt Malassé, A. (2011b). Les équilibres bipèdes permanents, origine embryonnaire, morphogenèse, équilibre occluso-postural, conséquences pour l'évolution psychomotrice et comportementale des Hominidés. HDR, Université de Technologie de Compiègne, Compiègne.

Dambricourt Malassé, A. (2013). Emergence of gracile human anatomy during the Holocene in Asia: Convergent acclimation and\or phylogenetic trend? Example of the occlusion. In *Holocene: Perspectives, Environmental Dynamics and Impact Events*, Kotlia, B.H. (ed.). Nova Sciences Publishers, New York.

Dambricourt Malassé, A. (2014a). Des tarsiers de Montauban aux Homo erectus de Chine : l'apport de Pierre Teilhard de Chardin à la compréhension de nos origines. In *Centenaire des recherches de Pierre Teilhard de Chardin sur les collections fossiles du Musée d'Histoire Naturelle Victor Brun*. Maison de la culture de Montauban, Montauban.

Dambricourt Malassé, A. (2014b). Le processus d'humanisation ou l'altérité territoire d'humanité. In *Interculturel, diaspora indienne et mondialisation : réalités, enjeux et perspectives, 2ème Colloque international de l'ODI-Réunion*, Saint Denis, Réunion, France, 29–30 April.

Dambricourt Malassé, A. (2016). The first Indo-French Prehistorical Mission in Siwaliks and the discovery of anthropic activities at 2.6 million years. *Comptes Rendus Palevol, Special Issue: Human Origins in the Indian Sub-continent*, 15(3–4), 281–294.

Dambricourt Malassé, A. (2017). The last threshold. *Inference, International Review of Science*, 3, 1. [Online]. Available at: https://inference-review.com/article/the-last-threshold.

Dambricourt Malassé, A. (2019). Hominin activities in the sub-Himalayan floodplain during the late Pliocene. In *Who Was Who, and Who Did What, Where and When?* Coppens, Y. and Vialet, A. (eds). Pontifical Academy of Sciences and CNRS, Vatican City.

Dambricourt Malassé, A. (2021a). Activités hominiennes dans la plaine inondable sous-himalayenne à la fin du Pliocène. In *Who Was Who, and Who Did What, Where and When?* Coppens, Y. and Vialet, A. (eds). CNRS, Editions, Paris.

Dambricourt Malassé, A. (2021b). The strengthening of the central nervous system in the human lineage and the emergence of shamanism. In *Anthropomorphism in Material Culture and Landscape: Approaching a Fundamental of Human Cognition*, Gheorghiu, D. and Paladino, V. (eds). Archaeopress, Oxford.

Dambricourt Malassé, A. (2022). The erect posture of Homo sapiens: The last threshold of a self-organized embryo-phylogenetic process with self-memorizing properties. In *Self-Organization as a New Paradigm in Evolutionary Biology: From Theory to Applied Cases in the Tree of Life*, Dambricourt Malassé, A. (ed.). Springer-Nature, Basingstoke.

Dambricourt Malassé, A. and Deshayes, M.J. (1990). Relationships between neuro-cephalic development and cranio-facial morphogenesis in living and fossil primates. *Fysisch-Anthropologische Mededelingen*, 2, 3–6.

Dambricourt Malassé, A. and Deshayes, M.J. (1992). Modeling of the craniofacial architecture during ontogenesis and phylogenesis. In *The Head-Neck Sensory Motor System*, Berthoz, A., Vidal, P.P., Graf, W. (eds). Oxford University Press, New York/Oxford.

Dambricourt Malassé, A. and Deshayes, M.J. (1995). Données ontogéniques et fonctionnelles des capacités langagières néandertaliennes. Implications pour les interprétations paléo-éthologiques. In *Nature et culture : actes du colloque international de Liège, 13–17 décembre*, Otte, M. (ed.). Univeristé de Liège, Liège.

Dambricourt Malassé, A. and Lallouet, F. (2009). Étude au scanner des canaux semi-circulaires d'un crâne sec d'anencéphale dépourvu de flexion sphéno-occipitale ; implications pour l'interprétation paléo-embryologique des Hominidés fossiles. *Avancées en biométrie humaine, Biométrie Humaine et Anthropologie*, 27(1–2), 79–86.

Dambricourt Malassé, A., Deshayes, M.J., Magnies-Jannin, F., Chaline, J., Marchand, D. (1999). A solution to human paradox: Fundamental ontogenies and heterochronies. *Human Evolution*, 14(4), 277–300.

Dambricourt Malassé, A., Martin, J.P., de Kerviler, E. (2000a). Approche biodynamique des angulations de la base du crâne de l'embryogenèse à la morphogenèse postnatale. In *L'identité humaine en question. Nouvelles problématiques et nouvelles technologies en paléontologie humaine et en paléoanthropologie biologique*, Andrieux, P., Hadjouis, D., Dambricourt Malassé, A. (eds). Artcom, Paris.

Dambricourt Malassé, A., Martin, J.P., de Kerviler, E. (2000b). Neural tube, spheno-occipital flexion and semi-circular canals in fossil and modern hominids. *Acta Anthropologica Sinica*, 19, 82–89.

Dambricourt Malassé, A., Raj, R., Shah, S. (2015). Orsang Man: A robust *Homo sapiens* in Central India with Asian *Homo erectus* features. In *Recent Discoveries and Perspectives in Human Evolution, International Series 2719*, Sankhyan, A.R. (ed.). British Archaeological Record, Oxford.

Dambricourt Malassé, A., Singh, M., Karir B., Gaillard, C., Bhardwaj, V., Moigne, A.M., Abdessadok, S., Chapon Sao, C., Gargani, J., Tudryn, A., Calligaro, T., Kaur, A., Pal, S., Hazarika, M. (2016a). Anthropic activities in the fossiliferous Quranwala Zone, 2.6 Ma, Siwaliks of Northwest India, historical context of the discovery and scientific investigations. *Comptes Rendus Palevol, Special Issue: Human Origins in the Indian Sub-continent*, 15(3–4), 295–316.

Dambricourt Malassé, A., Moigne, A.M., Singh, M., Calligaro, T., Karir, B., Gaillard, C., Kaur, A., Bhardwaj, V., Pal, S., Abdessadok, S., Chapon Sao, C., Gargani, J., Tudryn, A., Garcia Sanz, M. (2016b). Intentional cutmarks on bovid from the Quranwala Zone, 2.6 Ma, Siwalik Frontal Range, northwestern India. *Comptes Rendus Palevol, Special Issue: Human Origins in the Indian Sub-continent*, 15(3–4), 317–339.

Dambricourt Malassé, A., Zhang, P., Wils, P. (2018). A new molar in the Middle Pleistocene hominid assemblage of Yanhuidong, Tongzi, South China. *Acta Anthropologica*, 37(1), 1–17.

Dawson, C. and Smith Woodward, A. (1913). On the discovery of a Palaeolithic human skull and mandible in a flint-bearing gravel overlying the Wealden (Hastings beds) at Piltdown, Fletching (Sussex). *Quarterly Journal of the Geological Society of London*, 69, 117–123.

Dean, M.C. and Wood, B.A. (1982). Basiskull anatomy of Plio–Pleistocene hominids from East and South Africa. *Am. Journ. Phys. Anthropol.*, 59(2), 157–174.

Dechambre, E. (1948–1949). L'évolution des primates et la théorie de la fœtalisation. *Mammalia*, 12, 27–41; 13, 100–117.

Delaire, J. (2006). Le concept "cortical". *International Orthodontics*, 4, 241–260.

Delattre, A. and Fénart, R. (1960). *L'hominisation du crâne étudiée par la méthode vestibulaire*. CNRS, Paris.

Deloison, Y. (2004). A new hypothesis on the origin of Hominoid locomotion. In *From Biped to Strider: The Emergence of Modern Human Walking, Running, and Resources Transport*, Meldrum, J. and Hilton, C.E. (eds). Kluwer Academic/Plenum Publishers, New York.

Deshayes, M.J. (1986). *Croissance cranio-faciale et orthodontie*. Masson, Paris.

Deshayes, M.J. (1991). Reconsidération de la croissance crânio-faciale au cours du développement et de l'évolution. Conséquences en orthopédie dento-maxillo-faciale. *Rev. Orthop. Dento Faciale*, 25, 353–365.

Deshayes, M.J. and Dambricourt Malassé, A. (1991). Analyse des différents types architecturaux cranio-faciaux par l'approche ontogénique de l'hominisation. *Rev. Stom. ChirMaxfac.*, 4, 249–258.

Deshayes, M.J., Dambricourt Malassé, A., Deshayes, P. (1990). Utilisation de la téléradiographie dans la compréhension de la mécanique humanisante. In *L'image et la Science. Proceedings of the 115th National Congress of Learned Societies*, 149–157.

Dubois, E. (1894). Pithecanthropus erectus, eine menschenähnliche Übergangsform aus Java. Landesdruckerei, Batavia.

Dubois, E. (1896). Le Pithecanthropus erectus et l'origine de l'homme. *Bulletins et Mémoires de la Société d'Anthropologie de Paris*, 7, 460–467.

Dubois, E. (1897). Sur le rapport du poids de l'encéphale avec la grandeur du corps chez les mammifères. *Bulletins et Mémoires de la Société d'Anthropologie de Paris*, 8, 337–376.

Edinger, T. (1948). Evolution of the horse brain. *Geological Society of America Memoir*, 25, 187.

Etler, D. (2010). It is time to sink the genus Australopithecus. *Sinanthropus website. Anthropology Department, Cabrillo College* [Online]. Available at: http://sinanthropus.blogspot.com/2010/08/its-time-to-sink-genus-australopithecus.html.

Ferguson, W.W. (1984). Revision of fossil hominid jaws from the Plio/Pleistocene of Hadar in Ethiopia, including a new species of the genus Homo (Hominoidea: Homininae). *Primates*, 25, 519–529.

Fischer, E. (1902). Zur Kenntnis des Primordialcraniums der Affen. *Anat. Anz.*, 20, 410–417

Fleury, V. (2011). Dynamic topology of cephalochordate to amniote morphological transition: A self-organized system of Russian dolls. *C. R. Acad. Sci.*, 334(2), 91–99.

Gardiner, J. (2013). Evolutionary basins of attraction and convergence in plants and animals. *Communicative and Integrative Biology*, 6(6), e26760. [Online]. Available at: https://www.ncbi.nlm.nih.gov/pmc/articles/PMC3914912/.

Gehring, W.J. (1999). *La drosophile aux yeux rouges : gènes et développement*. Odile Jacob, Paris.

Godard, B. and Ducasse D. (1973). L'examen radiologique en O.D.F. L'analyse de Gudin. *Rev. Orthop. Dento-Faciale*, 7(3), 373–392.

Gould, S. (1980). *The Panda's Thumb, More Reflections in Natural History*. Norton, New York.

Gribnau, A. and Geijsberts, G. (1981). *Developmental Stages in the Rhesus Monkey (Macaca mulatta). Advances in Anatomy, Embryology and Cell Biology*. Springer-Verlag, Berlin, Heidelberg, New York.

Grimaldi, A. (1906). Discours d'ouverture [Online]. Available at: www.area-archives.org/area-archives/virtual-exhibition-fr/iph-musee.htm.

Grover, S.K. (1962). Sphenoidal angle and brain size. *American Anthropologist*, New Series, 64(3), Part 1, 521–523.

Gudin, R. (1952). Étude du crâne de profil. Morphologie et radiographie. *Bulletins et Mémoires de la Société d'Anthropologie de Paris*, 3(10), 59–87.

Guimberteau, J-C. and Armstrong, C. (2016). *L'architecture du corps humain vivant. Le monde extracellulaire, les cellules et le fascia révélés par l'endoscopie intratissulaire*. Sully, Vannes.

Gunz, P., Neubauer, S., Maureille, B., Hublin, J.-J. (2011). Virtual reconstruction of the Le Moustier 2 newborn skull. Implications for Neandertal ontogeny. *Paleo*, 22, 155–172.

Hadjouis, D. (1999). *Les populations médiévales du Val-de-Marne. Dysharmonies cranio-faciales, maladies bucco-dentaires et anomalies du développement dentaire au cours du Moyen Age*. Artcom, Paris.

Hadjouis, D. (2015). *Camille Arambourg ; un paléontologue, de l'Algérie à l'Afrique profonde*. L'Harmattan, Paris.

Hadjouis, D. (2018). *Atlas des maladies et traumatismes du monde médiéval et moderne*. ISTE Editions, London.

Hadjouis, D. (2021). *The Skull of Quadruped and Bipedal Vertebrates: Variations, Abnormalities and Joint Pathologies*. ISTE Ltd, London and John Wiley & Sons, New York.

Hague, R. (1966). *Man's Place in Nature: The Human Zoological Group*. Collins, Fontana Books, London.

Haile-Selassie, Y., Suwa, G., White, T.D. (2004). Late Miocene teeth from Middle Awash, Ethiopia, and early hominid dental evolution. *Science*, 303(5663), 1503–1505.

Haile-Selassie, Y., Latimer, B.M., Alene, M., Deino, A.L., Gibert, L., Melillo., S.M., Saylor, B.Z., Scott, G.R., Lovejoy, C.O. (2010). An early *Australopithecus* afarensis postcranium from Woranso-Mille, Ethiopia. *Proc. Natl. Acad. Sci. USA*, 107(27), 12121–12126.

Haile-Selassie, Y., Melillo S., Vazzana A., Ryan T.M. (2019). A 3.8-million-year-old hominin cranium from Woranso-Mille, Ethiopia. *Nature*, 573(7773), 1–6.

Haileab, B. and Brown, F.H. (1992). Turkana Basin-Middle Awash Valley correlations and the age of the Sagantole and Hadar Formations. *J. Hum. Evol.*, 22(6), 453–468.

Haldane, J.B.S. (1950). Les mécanismes de l'évolution. In *Paléontologie et transformisme*, Arambourg, C., Cuénot, L., Grassé, P.-P. (eds). Albin Michel, Paris.

Hammond, A.S., Rook, L., Anaya, A.D., Cioppi, E., Costeur, L., Moyà-Solà, S., Almécija, S. (2020). Insights into the lower torso in late Miocene hominoid *Oreopithecus bambolii*. *Proc. Natl. Acad. Sci. USA*, 117(1), 278–284.

Han, F., Bahain, J-J., Deng, C., Boëda E., Hou, Y., Wei, G., Huang, W., Garcia, T., Shao, Q., He, C., Falguères, C., Voinchet, P., Yin, G. (2015–2017). The earliest evidence of hominid settlement in China: Combined electron spin resonance and uranium series (ESR/U-series) dating of mammalian fossil teeth from Longgupo cave. *Quaternary International*, 434(A), 175–183.

Harmand, S., Lewis, J.E., Feibel, C.S., Lepre, C.J., Prat, S., Lenoble, A., Boës, X., Quinn, R.L., Brenet, M., Arroyo, A., Taylor, N., Clément, S., Daver, G., Brugal, J.-P., Leakey, R.L., Mortlock, A., Wright, J.D., Lokorodi, S., Kirwa, C., Kent, D.V., Roche, H. (2015). 3.3-million-year-old stone tools from Lomekwi 3, West Turkana, Kenya. *Nature*, 521, 310–315.

Harrisona, T., Xueping, J., Suc, D. (2002). On the systematic status of the late Neogene hominoids from Yunnan. *J. Hum. Evol.*, 43, 2, 207–227.

Haug, E. (1900). Les géosynclinaux et les aires continentale. *Bulletin de la Société géologique de France*, Series 3, 28, 617–711.

Heim, J.L. (1989). La nouvelle reconstitution du crâne néandertalien de La Chapelle-aux-Saints. Méthodes et résultats. *Bulletins et Mémoires de la Société d'Anthropologie de Paris*, 1(1–2), 95–117.

Herries, A.I.R., Pickering, R., Adams, J.W., Curnoe, D., Warr, G., Latham, A., Shaw, J. (2013). A multi-disciplinary perspective on the age of *Australopithecus* in southern Africa. In *Paleoecology of Australopithecus*, Leakey, L. and Reed, K.E. (eds). Springer-Verlag, New York.

Holloway, R.L. (1972). New australopithecine endocast, SK 1585, from Swartkrans, South Africa. *American Journal of Physical Anthropology*, 37, 173–186.

Hominid Fossil Repository (n.d.). *Australopithecus anamensis*–KNM-KP 29285 [Online]. Available at: http://projects.leadr.msu.edu/hominidfossils/items/show/42.

Hublin, J.J., Spoor, F., Braun, M., Zonneveld, F., Condemi, S. (1996). A late Neanderthal from Arcy-sur-Cure associated with Upper Palaeolithic artefacts. *Nature*, 381, 224–226.

Hublin, J.J., Ben-Ncer, A., Bailey, S.E., Freidline, S.E., Neubauer, S., Skinner, M.M., Bergmann, I., Le Cabec, A., Benazzi, S., Harvati, K., Gunz, P. (2017). New fossils from Jebel Irhoud, Morocco and the pan-African origin of *Homo sapiens*. *Nature*, 546(7657), 289–292.

Hurel, A. (2011). *L'abbé Breuil. Un préhistorien dans le siècle*. CNRS Editions, Paris.

Jacobson, A.G. (1980). Computer modeling of morphogenesis. *American Zoologist*, 20(4), 669–677.

Jeffery, N. and Spoor, F. (2004). Prenatal growth and development of the modern human labyrinth. *J. Anat.*, 204, 71–92.

Jeffery, N., Williams, S., Davies, K., Kockenberger, W. (2007). Craniofacial growth in fetal Tarsius bancanus: Brains, eyes and nasal septa. *J. Anat.*, 210, 703–722.

Ji, X., Jablonski, N., Su, D.F., Flynn, L., Deng, C., You, Y., Kelley, J. (2013). A new juvenile hominoid cranium from the terminal Miocene of Yunnan, China. *Chinese Science Bulletin*, 58, 3771–3779.

Johnson, T. (2005). *Homo sapiens, une nouvelle histoire de l'Homme*. Arte, Discovery Communication Inc., NMO, Paris.

Kimbel, W.H. and Rak, Y. (1993). The importance of species in paleoanthropology and an argument for the phylogenetic concept of the species category. In *Species, Species Concepts, and Primate Evolution*, Kimbel, W.H. and Martin, L.B. (eds). Plenum Press, New York.

Kimbel, W.H. and Rak, Y. (2010). The cranial base of *Australopithecus* afarensis: New insights from the female skull. *Philos. Trans. R. Soc. Lond., B. Biol. Sci.*, 365(1556), 3365–3376.

Langman, J. (1984). *Abrégé d'embryologie médicale.* Masson, Paris.

Leroi-Gourhan, A. (1983). *Mécanique vivante, le crâne des vertébrés du poisson à l'homme.* Fayard, Paris.

Levi, G. (1900). Beitrag zum studium der entwickelung des knorpeligen primordialcraniums des menschen. *Arch. Mikr. Anat.*, 55, 341–414.

Lewis, W.H. (1920). The cartilaginous skull of a human embryo twenty-one millimeters in length. *Carnegie Contributions to Embryology*, 272(9), 299–324.

Lieberman, D.E. and McCarthy, R.C. (1999). The ontogeny of cranial base angulation in humans and chimpanzees and its implications for reconstructing pharyngeal dimensions. *J. Hum. Evol.*, 36, 487–517.

de Lumley, M.A. (1976). Les Anténéandertaliens dans le Sud. In *La Préhistoire française : les civilisations paléolithiques et mésolithiques*, de Lumley, H. (ed.). CNRS, Paris.

de Lumley, H. (2007). *La grande histoire des premiers hommes européens.* Odile Jacob, Paris.

de Lumley, M.A. and Sonakia, A. (1985). Premiere découverte d'un *Homo erectus* sur le continent indien á Hathnora, dans la moyenne vallée de la Narmada. *L'Anthropologie*, 89(1), 13–61.

Macklin, C.C. (1914). The skull of a human fetus of 40 mm. *American Journal of Anatomy*, 16(3), 317–385.

Manouvrier, L. (1895). Discussion sur le "*Pithecanthropus erectus*" comme précurseur présumé de l'homme. *Bulletins et Mémoires de la Société d'Anthropologie de Paris*, 6, 12–47.

Manouvrier, L. (1896). Le *Pithecanthropus erectus* et la théorie transformiste. *Revue scientifique*, 4(5), 289–299.

McCown, T.D. and Keith, A. (1939). *The Stone Age of Mount Carmel. The Fossil Human Remains from the Levalloiso-Mousterian.* Oxford University Press, Oxford.

McPherron, S.P., Alemseged, Z., Marean, C.W., Wynn, J.G., Reed, D., Geraads, D., Bobe, R., Béarat, H.A. (2010). Evidence for stone-tool-assisted consumption of animal tissues before 3.39 million years ago at Dikika, Ethiopia. *Nature*, 466, 857–860.

Mitsuru, M., Takahisa, S., Tatsuyoshi, S., Shinkichi, N. (2002). Biomechanical influences of head posture on occlusion, an experimental study using finite elements analysis. *The European Journal of Orthodontics*, 24(4), 319–326.

Morin, E. (2001). *La méthode 5. L'humanité de l'humanité – L'identité humaine*. Le Seuil, Paris.

Morse, P.E., Chester, S.G.B., Boyer, D.M., Smith, T., Smith, R., Gigase, P., Bloch, J.I. (2019). New fossils, systematics, and biogeography of the oldest known crown primate Teilhardina from the earliest Eocene of Asia, Europe, and North America. *J. Hum. Evol.*, 128, 103–131.

Mounier, A. and Mirazón Lahr, M. (2019). Deciphering African late middle Pleistocene hominin diversity and the origin of our species. *Nature Communications*, 10(1), 1–13.

Müller, F. and O'Rahilly, R. (1980a). The early development of the nervous system in staged insectivore and primate embryos. *J. Comp. Neurol.*, 193, 741–751.

Müller, F. and O'Rahilly, R. (1980b). The human chondrocranium at the end of the embryonic period, proper, with particular reference to the nervous system. *Am. J. Anat.*, 159, 33–58.

Ni, X.-J., Wang, Y.-Q., Hu, Y.-M., Li, C.-K. (2004). A euprimate skull from the early Eocene of China. *Nature*, 427, 65–68.

Nusse, H.E., Yorke, J.A., Kostelich, E.J. (1994). Basins of attraction. In *Dynamics: Numerical Explorations*. Springer, New York.

Olivier, G. (1965). *Anatomie anthropologique*. Vigot Frères, Paris.

O'Rahilly, R. (1975). *A Color Atlas of Human Embryology*. Sounders, Philadelphia, PA.

O'Rahilly, R. (1981). Introduction à l'étude des stades embryonnaires chez l'Homme. *Bull. Asso. Anat.*, 65(189), 141–236.

O'Rahilly, R., Müller, F., Meyer, D.B. (1980). The human vertebral column at the end of the embryonic period proper. 1. The column as a whole. *J. Anat.*, 131, 565–575.

Paidas, C.N., Moreale, R.F., Holoski, K.M., Lund, R.E., Hutchins, G.M. (1999). Septation and differentiation of the embryonic human cloaca. *J. Ped. Surg.*, 34, 877–884.

Pickford, M., Senut, B., Gommery, D., Treil, J. (2002). Bipedalism in Orrorin tugenensis revealed by its femora. *Comptes Rendus Palevol*, 1(4), 191–203.

Piveteau, J. (1950a). Le débat entre Cuvier et Geoffroy Saint-Hilaire sur l'unité de plan et de composition. *Revue d'histoire des sciences*, 3–4, 343–363.

Piveteau, J. (1950b). Paléontologie, génétique et transformisme. In *Paléontologie et transformisme*, Arambourg, C., Cuénot, L., Grassé, P.-P. (eds). Albin Michel, Paris.

Piveteau, J. (1963). *Des premiers vertébrés à l'Homme*. Albin Michel, Paris.

Piveteau, J. (1969). Préface de H. de Lumley-Woodyear. In *Le Paléolithique inférieur et moyen du Midi Méditerranéen dans son cadre géologique*, de Lumley-Woodyear, H. (ed). Gallia-Préhistoire, CNRS, Paris.

Ponce de León, M.S., Lubov, G., Doronichev, V., Romanova, G., Akazawa, T., Kondo, O., Ishida, H., Zollikofer, C. (2008). Neanderthal brain size at birth provides insights into the evolution of human life history. *Proc. Natl. Acad. Sci. USA*, 5(37), 13764–13768.

Prenant, M. (1950). Commentaires à "Mécanismes de l'évolution". In *Paléontologie et transformisme*, Arambourg, C., Cuénot, L., Grassé, P.-P. (eds). Albin Michel, Paris.

Prigogine, I. and Stengers, I. (1979). *La Nouvelle Alliance. Métamorphose de la science.* Gallimard, Paris.

Roth, H. (1983). Comparaison statistique de la forme des arcades alvéolaires et dentaires des mandibules fossiles. Position de l'Homme de Tautavel dans l'évolution humaine. PhD Thesis, Université Marseille 1, Marseille.

Saban, R. (1995). *Aux sources du langage articulé.* Masson, Paris.

Schmahmann, J. (1991). An emerging concept: The cerebellar contribution to higher function. *Archives of Neurology*, 48, 1178–1187.

Schoenwolf, G.C., Bleyl, S., Brauer, P. Francis-West, P. (2014). *Larsen's Human Embryology*, 5th edition. Churchill Livingstone, London.

Science (2009). *Ardipithecus ramidus. AAAS*, Special edition [Online]. Available at: https://www.sciencemag.org/site/feature/misc/webfeat/ardipithecus/index.xhtml.

Senut, B. (2009). L'évolution de l'homme. In *Le rift est-africain*, Hirsch, B. and Roussel, B. (eds). IRD Éditions, Publications scientifiques du MNHN, Marseille [Online]. Available at: https://books.openedition.org/irdeditions/1744?lang=fr.

Simons, E.L., Seiffert, E.R., Ryan, T.M., Attia, Y. (2007). A remarkable female cranium of the early Oligocene anthropoid *Aegyptopithecus zeuxis* (Catarrhini, Propliopithecidae). *Proc. Natl. Acad. Sci. USA*, 104(21), 8731–8736.

Simpson, G.G. (1950). L'orthogenèse et la théorie synthétique de l'évolution. In *Paléontologie et transformisme*, Arambourg, C., Cuénot, L., Grassé, P.-P. (eds). Albin Michel, Paris.

Smith, B.R. (1999). Visualizing human embryo. *Scientific American*, 280(3), 76–81.

Spoor, F. (1997). Basicranial architecture and relative brain size of Sts 5 (*Australopithecus* africanus) and other Plio-Pleistocene hominids. *South African Journal of Science*, 93, 182–186.

Spoor, F. and Zonneveld, F. (1997). The bony labyrinth of Sts 5 ("Mrs Ples"). *South African Journal of Science*, 93, 180.

Spoor, F. and Zonneveld, F. (1998). A comparative review of human bony labyrinth. *American Journal of Physical Anthropology*, 41, 211–251.

Spoor, F., Wood, B., Zonneveld, F. (1994). Implication of early hominid labyrinthine morphology for the evolution of human bipedal locomotion. *Nature*, 369, 645–648.

Spoor, F., Jeffery, N., Zonneveld, F. (2000). Imaging skeletal growth and evolution. In *Development, Growth and Evolution. Implications for the Study of the Hominid Skeleton*, O'Higgins, P. and Cohn, M. (eds). Linnean Society of London, London.

Spoor, F., Hublin, J.-J., Braun, M., Zonneveld, F. (2003). The bony labyrinth of Neanderthals. *J. Hum. Evol.*, 44(2), 141–165.

Spoor, F., Leakey, M.G., Leakey, L.N. (2010). Hominin diversity in the Middle Pliocene of eastern Africa: The maxilla of KNM-WT 40000. *Philos. Trans. R. Soc. Lond. B. Biol. Sci.*, 365(1556), 3377–3388.

Starck, D. (1960). Das Cranium eines Schimpansenfetus (Pan troglodytes Blumenbach 1799) von 71 mm Sch.-Stlg., nebst Bemerkungen über die Körperform von Schimpansenfeten. (Beitrag zur Kenntnis des Primatencraniums II). *Morph. Jb.*, 100, 559–647.

Susman, R.L. (1991). Who made the Oldowan tools? Fossil evidence for tool behavior in Plio-Pleistocene Hominids. *Journal of Anthropological Research*, 47(2), 129–151.

Suwa, G., Asfa, K., Reiko, T., Kubo, D., Lovejoy, C.O., White, T.D. (2009). The *Ardipithecus* ramidus skull and its implications for Hominid origins. *Science*, 326(5949), 68e1–68e7.

Suzuki, H. (1970). The skull of the Amud man. In *The Amud Man and His Cave Site*, Suzuki, H. and Takai, F. (eds). Keigaku Publishing, Tokyo.

Teilhard de Chardin, P. (1921a). La présence d'un Tarsier dans les Phosphorites du Quercy et sur l'origine tarsienne de l'Homme. *L'Anthropologie*, 31, 329–330.

Teilhard de Chardin, P. (1921b). Comment se pose aujourd'hui la question du Transformisme. *Études*, 167, 524–544.

Teilhard de Chardin, P. (1925). Le paradoxe transformiste. A propos de la dernière critique du transformisme par M. Vialleton. *Revue des questions scientifiques*, 7(1), 53–80.

Teilhard de Chardin, P. (1937). Jubilé de Marcellin Boule. *L'Anthropologie*, 47, 599–600.

Teilhard de Chardin, P. (1950a). Les Australopithèques et le chaînon manquant ou "missing link" de l'évolution. *Études*, June, 340–345.

Teilhard de Chardin, P. (1950b). Sur un cas remarquable d'un cas d'orthogenèse de groupe : l'évolution des Siphnéidés de Chine. In *Paléontologie et transformisme*, Arambourg, C., Cuénot, L., Grassé, P.-P. (eds). Albin Michel, Paris.

Teilhard de Chardin, P. (1955). *Le phénomène humain*. Le Seuil, Paris.

Teilhard de Chardin, P. (2004). *Correspondance inédite (1923–1940)*. Éditions du Muséum-Edisud, Paris.

Thomas, H. (2002). *Le mystère de l'homme de Piltdown. Une extraordinaire imposture scientifique*. Belin, Paris.

Villmoare, B., Kimbel, W.H., Seyoum, C., Campisano, C.J., DiMaggio, E., Rowan, J., Braun D.R., Arrowsmith, J., Reed, R., Kaye, E. (2015). Paleoanthropology. Early Homo at 2.8 Ma from Ledi-Geraru, Afar, Ethiopia. *Science*, 347(6228), 1352–1355.

Walker, A., Falk, D., Smith, R., Pickford, M. (1983). The skull of *Proconsul africanus*: Reconstruction and cranial capacity. *Nature*, 305, 525–527.

Ward, C., Leakey, M., Walker, A. (1999). The new hominid species *Australopithecus anamensis*. *Evolutionary Anthropology*, 7, 197–205.

Ward, C., Leakey, M.G., Walker, A. (2001). Morphology of *Australopithecus anamensi* from Kanapoi and Allia Bay, Kenya. *J. Hum. Evol.*, 41, 255–368.

Ward, C., Kimbel, W., Johanson, D. (2011). Complete fourth metatarsal and arches in the foot of *Australopithecus afarensis*. *Science*, 331(6018), 750–753.

Watson, D.M.S. (1950). Commentaires. In *Paléontologie et transformisme*, Arambourg, C., Cuénot, L., Grassé, P.-P. (eds). Albin Michel, Paris.

Weaver, A.H. (2005). Reciprocal evolution of the cerebellum and neocortex in fossil humans. *Proc. Natl. Acad. Sci. USA*, 102(10), 3576–3580.

Weidenreich, F. (1928). *Der Schädelfund von Weimar-Ehringsdorf*. Fischer-Verlag, Jena.

Weidenreich, F. (1939). Six lectures on Sinanthropus pekinensis and related problems. *Bull. Geol. Soc. China*, 19, 110.

Weidenreich, F. (1943). The skull of Sinanthropus pekinensis. A comparative study on a primitive hominid skull. *Paleontologia Sinica*, New series D, 10, 126.

Weidenreich, F. (1951). Morphology of Solo man. *Anthropological Papers of the American Museum of Natural History*, 43, 203–290.

White, D., (2007). *Evolution of the Brain: Neuroanatomy, Development and Paleontology*. Wadsworth Publishing Co Inc., Belmont.

White, T.D., Suwa, T.D., Asfaw, G. (1994). *Australopithecus* ramidus, a new species of early hominid from Aramis, Ethiopia. *Nature*, 371, 306–312.

Wind, J. and Zonneveld, F.W. (1989). Computed tomography of an *Australopithecus* skull (Mrs Ples): A new technique. *Naturwissenschaften*, 76, 325–327.

Woldegabriel, G., White, T.D., Suwa, G., Renne, P., de Heinzelin, J., Hart, W.K., Heiken, G. (1994). Ecological and temporal placement of early Pliocene hominids at Aramis, Ethiopia. *Nature*, 371(6495), 330–333.

Wolpoff, M.H., Senut, B., Pickford, M., Hawks, J. (2002). Sahelanthropus or "Sahelpithecus"? *Nature*, 419, 581–582.

Wolpoff, M.H., Hawks, J., Senut, B., Pickford, M., Ahern, J. (2006). An ape or the ape: Is the Toumaï cranium TM 266 a hominid? *PaleoAnthropology*, 36–50.

Wong, M.D., van Eede, M.C., Spring, S., Jevtic, S., Boughner, J.C., Lerch, J.P., Henkelman, R.M. (2015). 4D atlas of the mouse embryo for precise morphological staging. *Development*, 142, 3583–3591.

Wood, B. and Constantino, P. (2007). *Paranthropus* boisei: Fifty years of evidence and analysis. *Yearbook of Physical Anthropology*, 50, 106–132.

Wu, X.-J., Crevecoeur, I., Liu, W., Xing, S., Trinkaus, E. (2014). Temporal labyrinths of eastern Eurasian Pleistocene humans. *Proc. Natl. Acad. Sci. USA*, 111(29), 10509–10513.

Xiaoyou, Y., Norman, J., Barlow, M., Feuston, H. (2017). Micro-computed tomography and volumetric imaging in developmental toxicology. In *Reproductive and Developmental Toxicology*, Gupta, R.C. (ed.). Elsevier, Paris.

Xu, Q. and Lu, Q. (2008). Lufengpithecus lufengensis, *an Early Member of Hominidae*. Science Press, Beijing.

Yue, L.P., Zhang, Y.X., Qi, G.Q., Heller, F.A., Wang, J.Q., Yang, L.R., Zhang, R. (2004). Paleomagnetic age and palaeobiological significance of hominoid fossil strata of Yuanmou Basin in Yunnan. *Science in China, Series D – Earth Sciences*, 47(5), 405–411.

Zhao, L., Lu, Q., Zhang, W. (2008). Age at first molar emergence in *Lufengpithecus lufengensis* and its implications for life-history evolution. *J. Hum. Evol.*, 54(2), 251–257.

Index

I

J, K

L

M

N

O

P

Q, R, S

T

V, W, Z

Other titles from

ISTE

in

Biology

2021

DAMBRICOURT MALASSÉ Anne
Embryogeny and Phylogeny of the Human Posture 1 (Comparative Anatomy and Posture of Animal and Human Set – Volume 3)

GRANDCOLAS Philippe, MAUREL Marie-Christine
Systematics and the Exploration of Life

HADJOUIS Djillali
The Skull of Quadruped and Bipedal Vertebrates: Variations, Abnormalities and Joint Pathologies (Comparative Anatomy and Posture of Animal and Human Set – Volume 2)

HULLÉ Maurice, VERNON Philippe
The Terrestrial Macroinvertebrates of the Sub-Antarctic Îles Kerguelen and Île de la Possession

2020

CAZEAU Cyrille
Foot Surgery Viewed Through the Prism of Comparative Anatomy: From

Normal to Useful (Comparative Anatomy and Posture of Animal and Human Set – Volume 1)

DUJON Bernard, PELLETIER Georges
Trajectories of Genetics

2019

BRAND Gérard
Discovering Odors

BUIS Roger
Biology and Mathematics: History and Challenges